UFO:
The making of a myth?

Aubecker Place Publications
U57 2 Hinder Street
Gungahlin ACT 2912
Australia
aubeckerr@gmail.com

Table of Contents

Dedication

This book is dedicated to anyone with an inquisitive mind interested in the subject of science and UFOs.

Foreword

This book, booklet, or novella – whatever you want to call it – discusses the study of UFOs from a scientific point of view. Therefore, it will introduce a brief history of what has transpired, touching upon landmark events or cases here and there. It critiques what has occurred in the fledgling field of 'Ufology' as people have tried to study the phenomenon. However, what it doesn't do is present an exhaustive treatise on UFOs. It also isn't an exhaustive discussion of individual cases or provide an exhaustive regurgitation of the history of UFOs. It doesn't interpret or re-interpret the history of UFOs, or what has occurred in the UFO field. And the term 'field of study' is used loosely. For an exhaustive recount of cases and events in this field, read Richard Dolan's *UFOs for the 21st-century mind*. Or any number of books dedicated to providing a history of UFOs. By taking a scientific point-of-view, however, this book highlights the approaches to the study of UFOs, the methods used, and their research methodologies. It is, therefore, a critical examination of the study of UFOs and doesn't try to be a defence of UFOs, in the sense that they are extraterrestrial or not. But it does try to encourage further research into the phenomenon using a scientific approach, using scientific methods and methodologies, and suggests ways to do this, once the dross of pseudoscience and mysticism encasing the field have been removed.

Prologue

"[to] promote research on UFOs to discover the true nature of
the phenomenon."
– the Mutual UFO Network (MUFON) 'Mission Statement
and Goals', Point II

1. Introduction

Epistemology: "the theory of knowledge, especially with regard to its methods, validity, and scope, and the distinction between justified belief and opinion."

There's a thirty-year-old man in a car, perhaps a 1960s Buick. He's an architect named David Vincent. He's driving in the small hours on an open, lonely road, returning from a meeting in a big city somewhere in the United States. The time is around four AM and he's looking for a diner with coffee and a place to rest. He soon pulls off the main road into 'Bud's Diner' but discovers it's closed. So he parks the car facing away from the diner, toward an open space almost the size of a small football field. He makes himself comfortable and tries to get some shuteye. But he is soon roused by a noise and a pulsing light. He wonders what the hell is going on and looks in the direction of the noise, blinks a few times to adjust his vision to the pulsing light and sees something he's never seen before. A saucer-shaped craft descending and landing perhaps fifty metres away on the open patch of land. The craft is about forty, perhaps fifty metres in diameter, and is glowing reds and greens and yellows. As he stares at the craft, he soon thinks he knows what he is seeing: a UFO. A craft from another world.

This is the opening scene to the first episode of *The Invaders*, a Quinn-Martin Production, made in the mid-1960s. It first appeared on TV in January 1967 and stars Roy Thinnes as David Vincent. From this day, David Vincent's life is turned upside down as he tries to convince people that he's seen an extraterrestrial craft, and that they are not here to make friends.

But he meets opposition, cynicism, and scepticism with most people, including his friends, who all think he's nuts – a loony-tunes. However, he knows what he saw, and is determined to tell the world. He goes in search of evidence to convince people that the world is being invaded by 'aliens'. But for some odd reason, the aliens never kill him, though they do try to stop him from getting the evidence he seeks. It's a tragedy because David Vincent never achieves his goal.

When people talk of UFOs, what in the world are they talking about? Some say they are talking about secret military weapons or military planes. Some, energy. Some, demons. Others – the majority – think they are extraterrestrial craft. Whatever they are talking about, one thing is sure, they are objects that cannot be readily or easily identified. This book will discuss UFOs as real objects and not things that aren't real, and, therefore, will ignore notions of energy, demons, and sprites as UFOs. To talk of phantoms, sprites, and demons is to talk of flights of fancy. It will touch on such things here and there, but is concerned with real objects – objects that are often seen by multiple witnesses and show up on radar equipment. The kinds of objects that caused governments of the world to investigate them, secretly if not openly. For governments to investigate them means they were a concern. If UFOs were ghosts, governments would not have dedicated time and energy to the issue. This book will not do what consecutive governments have done, though, or what hundreds of books have done on the subject, speculate, but will simply present the facts. However, as will be seen, the facts do not speak for themselves (though some people think they do). And, after careful consideration, the facts do not result in a conclusion that seems reasonable to anyone. And highlights the problems associated with concluding that UFOs are extraterrestrial in origin.

Therefore, the question that must be asked above all others is this one: Why do people think UFOs are extraterrestrial? The assumptions in this question are 1) people see real objects, and 2) they think the objects are of extraterrestrial origin. In other words, the assumption is that UFOs are made by extraterrestrials and not made by humans. But the question that must be asked next is, how do these people know UFOs are extraterrestrial in origin? And this is where we meet a fork in the road. There are problems associated with the claim that UFOs are extraterrestrial craft. Can you guess what they might be? The problem of talking about UFOs is the problem of talking about extraterrestrials. No, it's not that people talk about UFOs as real objects, because that is not the (real) problem; the problem is the type of real object that is seen is the problem. And, therefore, the problem of their origin.

The question we must now answer above all others is this one: Are UFOs extraterrestrial craft, as some believe? To answer this question will be the quest of this book.

To answer this question, we must understand the type of people who see UFOs and how they identify, classify, and categorise UFOs. People do a range of things when they see something strange. One of them is to compare the seen thing with everything they know that is similar, familiar, and then conclude that it is one of these. If not, then it is something they have never seen before (and they must construct a new category and classification system). If it is something they've never seen before, then it cannot be common. If it isn't common, then it is unusual. If it is unusual, it cannot be something everyone would ordinarily see. If it isn't something ordinarily seen, then it is not something anyone has seen. And therefore, the sight of the thing is unique (to them). Perhaps meant just for them. Or they see themselves as special because they have seen something no-one else has.

Their reaction is initially a psychological one. And everyone reacts in one way or another when they see something strange, unusual, and mystifying. But how they react can be placed on a continuum, from outright denial to unwavering belief. But why is this important?

The Mutual UFO Network (MUFON), an organisation dedicated to investigating UFO sightings, has collected reports of UFOs since its inception in mid-1969. It claims that more than 90% (perhaps 95%) of all UFO sightings are misidentifications. This is significant. To say that most claims are misidentifications is to say that most people seeing something strange and unusual immediately think they see something from another world (or dimension), a catchphrase in UFO circles, and report it to a group like MUFON. Perhaps a small proportion of these sightings are imaginary objects or imagined objects, and that is another problem. If a small proportion are imaginary objects, then the rest are real objects, and thus we meet the first conundrum. Although hoaxes and misidentification have continued to keep pace with eyewitness sightings of real objects, the problem arises when an attempt is made to sort out the misidentifications from the real thing. This is a problem for eyewitnesses as much as it is for the investigators who classify and categorise. It is also a problem for the sceptical and mainstream science. But what should the people who see UFOs do to address this problem? Some say, nothing; others, believe –

you simply must believe.

There is another group of people (often disparaged) who really should do something. Scientists. Why should they get involved? They should be involved because, according to the Condon Committee (2017) and the Condon Report, the US government found the issue worthy of serious investigation. And because one of the most glaring problems living in an otherwise infinite universe is the problem of being alone. This was something Carl Sagan, the famous American astronomer, planetary scientist, cosmologist, astrophysicist, astrobiologist, science communicator, author, and professor, famously attempted to address, and which now seems to be taking centre stage in astrophysics as exoplanets are discovered with the potential to harbour intelligent life. But what should science do to support the study of unidentified flying objects? For the former, the scientist would say the eyewitness should learn to critically analyse their own experience; realise the limitations of their sensory experience. But for the latter, the observer, they should try to help eyewitnesses identify UFOs, and solve the mystery of what they are.

The scientists, on the other hand, would say that the tendency of the first group (eyewitnesses) is to ignore the cautioning of their own minds and dismiss any suggestion that what they saw was not what they believe they saw. The tendency of the second group (scientists), when asked (which isn't often), is to deny the existence of UFOs and extraterrestrials and the possibility that they may be visiting planet Earth. Both groups, therefore, have a problem. One group just wants to believe (without evidence) that they saw an extraterrestrial craft; the other just wants the first to show some credible evidence (not fuzzy videos, for example). Of course, the hard-line both groups have historically taken has softened over the years as the possibility of extraterrestrials has taken centre stage with the discovery of exoplanets – planets outside the solar system.

The new enthusiasm for sending people to Mars has fired peoples' imaginations, too.

When the acronym 'UFO' is used in this book, it will be used as the singular: unidentified flying object. When the acronym 'UFOs' is used, it will be pluralised – *s* is added to 'UFO', e.g., UFOs. This may seem trivial, but its importance will be shown as the book progresses.

The first problem we encounter when studying, investigating, or researching the UFO phenomenon, as it is sometimes called, is when the acronym is used to signify the 'off-worldness' or 'other-worldness' of the object(s) in question, UFO. It was originally coined to simply mean unidentified flying object – singular. This term is still useful today, even if it is used to refer to things flying in the sky that aren't easily identifiable, at first glance. Hence the proliferation of UFOs and UFO sightings. However, UFOs are not merely unidentifiable flying objects. They are also unknown flying machines or craft, if you will, to use a quaint term. To be sure, they are machines. Like planes. Some people use the acronym UFO instead of 'craft' to describe the object. To be sure, a UFO is a craft (e.g., drone, piloted craft), and it is often believed (but not yet known) to be owned and piloted by beings from another world. And now, perhaps, we can begin to see the problems (if not so readily) – at least, it is hoped the reader can see the problems; can see them – associated with the official and unofficial study of UFOs, or Ufology, as it is sometimes called, in which claims are made but with little evidence to back them.

The generic term Ufology (pronounced /u-fology/) will be used to refer to all unofficial and official study, research, and investigation of UFOs.

In the past, perhaps not so much now, when people spoke of UFOs, they were ridiculed. Some were harassed; others, murdered, says Austin (2016). How can we talk about UFOs without being ridiculed or worse, rejected as an irrational thinking human being?

This book will try to answer this question, too, as it discusses the many theories about UFOs, their nature, and their origin. These theories will be analysed for their authenticity and/or 'truth'. Examples will be presented, claiming to provide crucial 'evidence' for the reality of UFOs. The lack of scientific evidence supporting the many theories put forward to explain UFOs will also be discussed to show how difficult and complex the task of studying UFOs is. And why theorising can be downright unhelpful.

Of course, discussing UFOs immediately creates a dilemma for the book: to repeat what others have said, or try to be original? The approach used in the book to discuss UFOs will be to follow a traditional path – the scientific method. Rational and logical. The traditional path will be used because it has stood the test of time. It is reliable; the best way to study something; anything. But, as it turns out, problematic for the study of UFOs, and this limitation is acknowledged. But isn't this true of anything science studies? Using the methods of science to analyse the phenomenon of UFOs, however, will turn some people off, and attract criticism. It is the only suitable tool to study UFOs in a rational and logical way. And, therefore, foreshadows the problems surrounding the many theories put forward to explain UFOs, their nature, and origins.

2. What's in a name?

Everything is in a name. We humans, in order to navigate our way through reality, label everything. We categorise and classify – from foods to movies. It's rational and logical. It's a practice that dates back more than a few hundred-thousand years, if not further. This practice of labelling, classifying, and categorising is likely linked to our nature as human beings. But it is now strongly linked to our language system. It's more than likely the same for other animals (see Gibbon language, decoded). So when we talk about something or reference it, we try to give it a name; label it. Or use an existing name, a general or even a specific one. 'Thing' just doesn't cut it sometimes. But it's a start; using *thing* immediately tells us that it is different from us and anything else we know. When we use the word *thing*, we are referring to something, not nothing. There is no such thing as nothing, except in the dictionary definition of the word which means the absence of something, e.g., an empty box – there's nothing in the box. When we talk about unidentified flying objects or UFOs, though, the commonly used label is the acronym UFO, used by Dr. J. Allen Hynek, and most people ever since, according to *A History of UFO Sightings* (2017). Read Richard Dolan's (2014) *UFOs for the 21st Century Mind*, for more information.

The inherent problem with the acronym UFO and using it is *identification*. When someone sees something flying or hovering in the sky, they will immediately think of it as a UFO, because 1) they cannot immediately identify it, and 2), UFO is now part of common vocabulary. Most people choose to use UFO because it now has strong socio-cultural connections and has become part of the world vernacular. Hence the term unidentified flying object. The object is in the sky; it is flying. Of course, the word flying does not suggest hovering. But UFOs have been known to hover, according to the *List of Reported UFO Sightings* (2017). And that's an interesting characteristic of UFOs. They can fly like a plane and hover like a helicopter. They can also do things that conventional, meaning common or standard planes, cannot, like complete a high-speed 90-degree turn.

Thus, using our traditional linguistic skills, we are apt to label something if we do not know what it is, with the simplest classifier, even if we think we know what it is we see.

However, while we can label something unusual or unidentifiable, just one thing, we can also describe it. Describing is an intriguing aspect of the UFO phenomenon. Describing a UFO is an interesting practice because it highlights the problems of eyewitness accounts – that eyewitness accounts can sometimes be unreliable – nine-out-of-ten UFO sightings are misidentifications (see MUFON). But the proliferation of sightings is a confounding factor. Confounding because the number of sightings cannot be accounted for in a logical way.

The most significant aspect of eyewitness accounts is the consistent description of a UFO or UFOs. And why some people, in psychology and the 'hard' sciences, at least, are sometimes dismissive of eyewitness accounts of UFOs, says Shostak in 'UFOs: Flying Emotions', SETI, and Fuller (1969) in *Aliens in the Skies: The New UFO Battle of the Scientists*. What eyewitnesses describe when they see a UFO is something that is quite unconventional, meaning uncommon or non-standard, and, therefore, apparently, easy to dismiss. If dismiss is what is actually being done. If it isn't conventional, according to psychologists, says Fuller (1969), then it must be imaginary. This aspect will be discussed in more detail later. But now, I'll discuss how people describe a UFO.

What did it look like? This is a common question asked of eyewitnesses (see MUFON). It also helps us sort misidentification from identification, which will be explained in more detail later. One of the most common answers to the question 'What did you see?' is that the object was circular, like a saucer, as in cup and saucer. Hence the earliest description, according to *A History of Sightings* (2017), of a UFO as being a flying saucer. An upside-down saucer. This term was used to describe the object's shape. It wasn't used to describe its nature, according to Cheryl Costa (2017), in their article 'The truth is in the UFO shapes'. As reporting of UFOs has progressed (formally since the late-1940s), so have the descriptions. Reasonable, under the circumstances. UFOs are not like other flying machines, like planes, helicopters, and zeppelins. However, with the advent of the drone (military and recreational), the comparison becomes uncanny and much harder to disambiguate. Planes behave in a consistent way: they fly in straight lines or in large circles. Unless it is a military plane with combat ability, then they appear more agile – e.g., the F-22 Raptor. But they tend not to hover and spin (see Harrier Jet), which was to become a common characteristic of saucer-shaped UFOs.

There is always an exception to the rule, though, and other shapes have since been identified. Some of the newer shapes, according to Cheryl Costa (2017), include spheres and cigars. Why the shape has 'evolved' from saucer to sphere or cigar is still unknown; here we will use the term 'evolve' because it can be argued that there is a strong socio-cultural link. But it is also thought to be linked with the species of extraterrestrial associated with the craft. More on this later. But, once again, the consistency of reporting, as will be seen, indicates something far more common. Once is unusual, but regularly is a pattern. And, by the way, science is in the business of finding patterns. If patterns are commonplace in reports of UFOs, then it is harder to dismiss them as imaginary or will-o-the-wisp. Unless every sighting is the result of mass hallucination, of course. But one thing is sure, a pattern of behaviour, shape, and materialisation does not mean that UFOs are extraterrestrial in origin; it just means UFOs aren't imaginary. It takes more than supposition to prove UFOs are extraterrestrial in origin, though. And that's an important distinction that will be discussed later.

The consistent reports of UFOs, though, is disquieting. Disquieting for two main reasons. Firstly, more people claim to see UFOs, and that means the phenomenon has become 'natural' and cross-cultural. Natural meaning it is like the weather. Cross-cultural because more than one culture claims to see UFOs, according to MUFON. In fact, there isn't a single culture (at the time of writing) that hasn't seen a UFO – as described in this book – historically, allegedly, or recently (modern times – between the 1940s and now). Read *A History of UFO Sightings* (2017), and Richard Dolan (2014) for clarification. If UFOs are seen everywhere, globally, then UFOs are a global phenomenon rather than isolated to a specific subgroup of a single culture. For instance, fifty years ago, that subgroup was farmers in the Mid-west of the United States. The range and type of people seeing UFOs today are diverse, not just farmers. Just about every occupation known to us has a member in its ranks that has seen a UFO, and that is significant. Of what, is yet to be clarified.

Secondly, the increasing number of sightings globally means one of two things. Either large numbers of human beings are experiencing a psychological disorder on a vast scale, or humans everywhere are being misled. In psychological terms, according to Langmaid (2017), the mass sightings are a confounded hallucination. The problem with this is that it is a shared hallucination. How can so many people be experiencing a shared hallucination? Mass sightings reveal a pattern, though; that similar objects are being seen around the world in different places. Are all these people experiencing the same hallucination? Impossible. Well, highly improbable, to be sure. It is highly unusual that disparate peoples, separated by vast oceans and cultures, are seeing the same things. Unless, of course, they are seeing the same things. Seeing the same things means they are seeing real, not imaginary objects.

Are humans being hoodwinked? This is a possibility, yes. That possibility is offered here. But by whom? And how? If people are seeing real objects, however, ones that can spin and hover, perhaps they are seeing cleverly designed and engineered craft? Is it possible, if people are not hallucinating, that what they are seeing is the latest advances in technology? Well, if they are then they deserve an apology. If engineers of these craft have let eyewitnesses be maligned, and that was wrong. The engineers have remained silent while the press, the government, and psychologists, among others, have distorted the state of mind of the eyewitnesses. These groups have done it by allowing what they build to be publicly denied and dismissed as the result of fertile imaginations.

Funny, because such craft need fertile imaginations for them to be conceived in the first place, and skilled workers (engineers) to build them.

Keeping all this in mind, the discussion now turns to exploring some of the many theories circulating about UFOs, their nature, and their origin.

3. Are UFOs real?

The theory that UFOs are real is a key place to begin. As was mentioned in the Introduction, whenever someone says they saw a UFO they are immediately criticised, worse, told they are delusional. This is unhelpful, if not downright unfriendly. But then there is the milder criticism: 'Are you sure?' Both question the veracity of the sighting, and ask the question, 'How do you know you saw a UFO?'

To answer the question 'Are UFOs real?' we need to ask whether there is any evidence. As also mentioned in the Introduction, science searches for patterns in the chaos not 'washing machines,' as poor Dr. Arroway was accused of doing in Carl Sagan's *Contact*. It's a metaphor, and we'll leave it there. Suffice it to say, searching for patterns in the chaos does often seem like the scientist is listening to the noise of washing machines. Noise. Especially patterns in cosmological 'noise'. There is a parallel here, in that those who study the UFO phenomenon must search the 'noise' surrounding it for the 'truth'. Truth is placed in quotation marks because the truth is out there, as Mulder from the *X-Files* TV show would say.

So how does someone who wants to study the UFO phenomenon go about finding the 'evidence' that supports the reality of UFOs? The question is much easier to answer today than it was fifty years ago. Back in the 1950s, it resulted in some interesting attempts to study the phenomenon. However, the findings of early research are controversial. Controversial because the method of study used may have been faulty. Where this line of thought leads will be taken up later. One of the things a researcher must do is find out who said what, when, and where. And perhaps, how they said it. The 'Roswell incident' in Roswell, New Mexico, in 1947, resulted in the public announcement that a UFO had allegedly crashed. In fact, according to Marcell, Jr. (2008), the witnesses said, 'flying saucer'. Either way, it was printed in the newspapers of the day, which made it a public record that can be searched and reviewed.

The alleged UFO crash in Roswell, New Mexico, referred to as the Roswell incident, is part of the historical record. It is, therefore, one of high significance. Alleged because the event has yet to be substantiated. The high significance, though, relates to the type of eyewitnesses involved in the event. The word 'type' isn't used to undervalue other eyewitnesses, but to show just how significant it was at the time it occurred. The eyewitnesses were members of the US military. Some other eyewitnesses were involved but weren't military; they were police and civilians. The military people had jobs that gave them direct involvement in the event, or at least after the event – the military claimed it was something that belonged to them. They were also asked to clean up the mess left at the crash site, for instance. That meant they were privy to the details of the crash. They saw the object that had crashed.

Being at the crash site put the military people in a position few other humans have been in. They saw something that many people still haven't. That many people still think isn't real. That many eyewitnesses are maligned for speaking out about. They saw a 'flying machine' unlike any they were privileged to see operating in the military at that time. Sure, they may have misidentified the object; that's a possibility. They may have misidentified it, as some suggest – that what they actually saw was a weather balloon – a claim made as public knowledge of the event grew due to increased media coverage. The question is, how do military personnel misidentify an object they would ordinarily easily identify on any other occasion, given their training and experience?

If the 'evidence' offered by the military personnel that what they saw was unlike anything they had ever seen before or since, and it is rejected, then an explanation must be offered, a decent one, for why they would misidentify the object. How could they have misidentified the object? Well, they might have been drunk at the time. Were they? No. They were sober. They were on duty. We could say they were cognitively impaired at the time. Were they? There is no medical evidence available to suggest they were not. None of the military personnel who saw and handled the debris were asked by their superiors to undergo a medical examination for having said that the object was not a balloon. But they were asked not to talk about it.

Aside from other issues surrounding the authenticity of the event, here is a group of highly trained individuals in the military seeing something highly unusual or unconventional. They claim that what they saw was not a balloon. In fact, they claim it was a 'flying saucer'. They describe the object as being 'saucer-shaped'. They describe it as being metallic. They claim there was debris. They also claim there were bodies. The presence of bodies is the most startling and controversial aspect of this story and will be discussed later.

The problem with these claims is that the 'evidence' is not available for anyone to examine. The eyewitness testimony can be mulled over for centuries. Why the physical evidence isn't available for examination is a question that should be asked.

Why isn't the crash debris available for anyone to examine if it wasn't a 'flying saucer,' as the superior officer in charge of recovering the debris later claimed? If it was a balloon, why isn't anyone allowed to see it and satisfy their curiosity that it wasn't what the military personnel who saw it were claiming it wasn't? This is a highly intriguing situation. It is intriguing because the absence of debris and restricted access to the 'balloon' debris has resulted in a conspiracy theory to grow and a myth to be propagated.

To make things even more interesting, the son (Marcell, Jr., 2008) of one of the military personnel involved in the recovery of the 'balloon' claims his father had secretly kept a piece of it. The son remembers seeing the piece and can describe what it looked like. He says there were markings that resembled some kind of hieroglyphics. This is astounding. If true, it supplies another piece of the puzzle that UFOs are extraordinary objects, and, as some speculate, may not have Earth as their origin. Yet the piece, like the 'balloon' debris, is unavailable for analysis because it was confiscated by the superior officer in charge of recovering the debris.

This story gives us a glimpse into how humans may behave when they meet something strange, 'other-worldly', or top-secret military craft. The military, for one. They may very well try to keep it hidden from the public. They may want to keep it to themselves. Especially if it represents technology that may assist them in becoming technologically superior to other countries on this planet. But that is pure speculation, and that is not the point of this book.

How can we sort out truth from conjecture? How can we sift decades of distortion and find ultimate truth? Something

UFO: THE MAKING OF A MYTH?

Robert Langdon of *The Da Vinci Code* might ask. Or is there no such thing? Are we barking up the wrong tree, as they say? Well, if this was an isolated event, we might well agree with that sentiment. But it wasn't. The Roswell incident, as it has come to be known, was not an isolated incident. There has been more 'noise' about UFOs than this particular incident has generated, and it has generated a lot of noise. Patterns have appeared where they may not have if it weren't for other similar incidents.

4. How else might we decide that UFOs are real?

Given that we cannot analyse the debris from the most famous UFO crash in history, allegedly, where else can we look to find 'evidence' to support the belief that UFOs are real? Well, it was said earlier that eyewitness accounts have continued since at least 1947 and is another place to find 'evidence'. In fact, according to *A History of UFO Sightings* (2017) and Leslie Kean (2011), some researchers have found that many UFO events occurred well before Roswell. Therefore, there were patterns of sightings well before Roswell. Has it continued unabated? Yes. How do we know? MUFON: The Mutual UFO Network. This group has done researchers and others a great service. MUFON has been collecting accounts of sightings since its inception in the late 1960s, and has been keeping records of UFO sightings ever since.

MUFON has collated 80,000 cases (at the time of writing). If one incident wasn't enough to convince people that UFOs are real, or, at least, real objects, then perhaps 80,000 more will suffice. Remember how earlier it was said that science looks for patterns in the chaos? Well, there's a pattern here. Since the late 1960s, or perhaps even earlier, people have been seeing UFOs, and many of them recorded in the MUFON CMS – case management system. But it's also conceivable that many haven't, and perhaps for obvious reasons: no one likes to talk about their experience for fear of ridicule. So it's conceivable that tens of thousands of sightings have gone unrecorded, and that is a conservative estimate, and a pity. But a pattern has emerged in the data. People have seen strange objects in the sky, not merely twinkling lights at night. Objects have been seen in the bright light of the day rather than fast-moving lights in the night sky.

If the thousands upon thousands of cases still don't convince you that people are seeing something unusual, if not strange, not the misidentifications of weather balloons, sunlight reflecting off plane windows and fuselage, flocks of birds, or swamp gas (just some of the explanations given by governments and 'authorised' military investigations into some sightings), then perhaps the eyewitness testimony of military and government employees will push your doubt over the edge into reality. Both the military and the governments of the world have been investigating UFOs. Official investigations of UFOs have been undertaken, and detailed reports published of the investigation and findings (see The Condon Report). Why bother investigating something that isn't real?

The question is, what would make the military and certain government departments (e.g., the CIA) officially investigate a phenomenon that has been ridiculed by psychologists and the press? Not to forget the simply sceptical. Why bother if it's all just a load of hoo-hah? A load of old cobblers? Why waste all that money and effort on something these particular groups of people would surely know, better than the rest of us, perhaps, is rubbish? It doesn't make sense that the military, for one, would investigate something that most psychologists seem to think is the result of some strange psychology. Some psychologists say that the experience of seeing a UFO can be attributed to a phenomenon called *false memory syndrome* and *sleep paralysis*. False memory syndrome is when the brain decoupages information from multiple sources into a memory, but a false one. Sleep paralysis is when a person is asleep and experiences a form of paralysis that results in hallucinations, e.g., strange beasts are in the room but the person is frozen and can't move. Why bother investigating something, using taxpayers' money, no less, if it's all happening in the heads of the eyewitnesses?

Even when the data has been looked at carefully, and all the misidentifications, hoaxes, and outright lies have been identified and removed, with the help of a great case management system, and excellent field investigators, UFOs were officially studied. During the 1960s, the US Air Force conducted an official investigation into UFOs. It called its investigation Project Blue Book (now a major TV show). Why would the US Air Force officially investigate something that couldn't possibly be real? Perhaps it had something to do with the fact that many military personnel were seeing UFOs. That many Air Force pilots, according to Leslie Kean (2011) and Richard Dolan (2014), were also seeing them. Military personnel saw them in WW1 and WW2, says Dolan (2014). They saw them in the skies over Vietnam and called them Foo Fighters.

At the height of the Cold War era, something strange was happening at nuclear missile sites in the United States. UFOs were appearing in the skies above the missile silos, and in one extraordinary incident, according to Richard Dolan (2014), UFOs were believed responsible for a shocking set of events. The UFOs switched off the missiles. Allegedly. There is an official record of this incident, and the officers involved are on the public record as having witnessed this extraordinary event. Some of the officers, according to the Disclosure Project's National Press Club briefing, are still alive today to tell their stories. Why would they risk ridicule and their reputations by telling a lie? How could something that is still reported in the literature as false memory syndrome and sleep paralysis switch off a nuclear missile? It doesn't make any sense.

The Introduction talks about two possibilities – if UFOs are real, that is. The first suggestion is that everyone who sees a UFO is delusional, and that includes large groups of ordinary people and expert eyewitnesses like military personnel, government agents, and police officers. These latter groups have been labelled 'expert' eyewitnesses because no one is more trained than they are in making judgements about what they see, and in their ability to describe the things they see. Yet there are sceptics. There are people who dismiss even the eyewitness testimony of such highly trained individuals as these. Who then is left to be believed when these eyewitnesses have been dismissed as delusional?

There is one other type of evidence that may yet convince. It is physical evidence. There is a type of evidence that is difficult to dismiss, and it is radar evidence. Still some people have and do try to dismiss it. Why would air traffic controllers fabricate evidence of unidentified flying objects? Wouldn't an air traffic controller be subject to investigation if they said their radar was showing an unidentified flying object in the skies above the airport? Well, many have and still do appear above airports or on the radar, according to Richard Dolan (2014), and others. There is radar evidence to support what airline pilots and others have seen in the skies around their aircraft, and pilots and personnel on the ground reporting what they have seen, say Leslie Kean (2011) and Richard Dolan (2014). Why would a pilot risk investigation by saying they saw an object near their plane that was unconventional? You would expect an airline pilot could tell the difference between another aircraft and a mirage, right?

The fact that civilian radar at international airports 'see' UFOs should make you wonder whether military radar also 'see' UFOs. The answer is, of course, yes. Military radar personnel have come forward in the Disclosure Project's National Press Club briefing, claiming the radar equipment they were working on showed UFOs. This type of evidence is hard to dismiss. Why would a military radar operator make such a claim? For money? Possibly. To get booted out of the army? Highly unlikely. To be considered a crank? Again, highly unlikely.

The US government, among many world governments, officially investigated the UFO phenomenon. Why would the US government invest money in a study of something that wasn't real? It seems the U.S. government thought them real enough between 1968 and 1969, because it financed a University of Colorado study of UFOs. Its final report was called *The Condon Report*. While the final report was suggestive, the press, when told about the investigation by the head investigator, Edward Condon, accepted his less-than-cheery summary of the UFO investigation rather than the report's actual findings. How odd. But there again, he was asked to investigate UFOs as a threat to national security and not UFOs as extraterrestrial craft. And this is a critical point.

In the Condon Report, Edward Condon concluded that UFOs were of no concern to the government and of no threat to US interests. But Leslie Kean (2011) thinks otherwise. This is a hugely different thing from concluding they don't exist. There was no mention of extraterrestrials because the study did not investigate the existence of extraterrestrials; merely whether UFOs were real and a threat. The study did not conclude that UFOs were not real, merely that they weren't a threat. And that is significant. Aside from what some UFO 're-searchers' say about this study – mostly negative things – it did what it set out to do; show that UFOs were real. If the Condon Report's conclusion bothers you, then you must conduct your own scientific study and produce a different outcome. Or, as they say in science research, replicate the study.

The tens of thousands of UFO cases in the MUFON case management system, the military, and government eyewitness testimony claim UFOs are real, and later official investigations of the phenomenon all support the conclusion that UFOs are real. That means that unidentified flying objects are real. But does it mean that UFOs are from somewhere else other than Earth and piloted by extraterrestrials? This is the sixty-four-million-dollar question and one without an answer, as yet.

5. Are UFOs extraterrestrial in origin?

This theory has been circulating for a long time. It started around the same time an object crashed in Roswell, New Mexico, in 1947. Perhaps the idea originated earlier. It has since become legendary. The only problem is that the 'evidence' is kept under lock and key, and no one can see it. Why would the military keep the evidence to themselves? What would be the advantage of doing so fifty-odd years after the event? For reasons of national security? Perhaps.

The theory that UFOs are real was discussed, and a massive amount of evidence was found to support it. So, the possibility that a UFO crashed in Roswell is strong, not weak. The claim of expert eyewitness testimony by military personnel – later denied by a high-ranking officer – to the crash is hard to dismiss without a good reason, and nobody can find one. Their claim that UFOs are real is supported by an official U.S. Air Force investigation called *Project Blue Book* and Dr. J. Allen Hynek's (1972) book *The UFO Experience: A Scientific Inquiry*. The U.S. Air Force had good reason to investigate UFOs because they were causing problems for planes and nuclear missiles. Two exceptionally good reasons to investigate the possibility that they are real and dangerous. Both Project Blue Book and The Condon Report stated that UFOs were perceived as a threat to national security. These investigations didn't consider whether they were real or not; they straight up investigated UFOs as a threat to national security. So bear that in mind as the discussion now focuses on whether UFOs are extraterrestrial in origin.

These investigations didn't conclude UFOs weren't real; they concluded they weren't a threat to national security. There is a difference here in meaning, and that is significant. The US Air Force and the US Government already knew UFOs were real. What they wanted to know was how much of a threat they were to national security. Why they concluded UFOs were not a threat to national security is puzzling. For one, military personnel claimed UFOs had shut down nuclear missiles in their silos. So why didn't that event constitute a threat?

The fact that the US Air Force and the US Government both knew UFOs were real is astonishing. It should also shock us. Why would these groups downplay the importance of this astounding revelation? There are two possibilities here. Firstly, they know what they are. They know what they are, and they don't see them as a threat. At least, not a threat to national security. Unless the rumours surrounding the UFO crash at Roswell are legitimate. One of the rumours concerns the reason for the crash. Why did the UFO crash? The story goes that the US military accidentally caused it to crash. It is said the crash had something to do with US radar technology. Fascinating.

Perhaps the reason why the U.S. military and the U.S. government do not see UFOs as a national security threat is because they know how to bring them down. Of course, what is lacking here is the hard evidence. All we have are rumours and conjecture. What we do know, for a fact, though, is that UFOs are real. The US Air Force and the US Government, and governments of other countries, according to UK UFO File Release and Nick Pope (2016), confirm that with their investigations. So the possibility that the rumour is true is also real.

Since writing this book, there have been astounding developments. The US Navy has released some videos of alleged UFOs recorded by Navy pilots. The other astounding development is that the US Government has ordered a report to be compiled of what it's military complex knows about UFOs. And even more astounding is that Congress has requested that UFOs be investigated using public money and departments be established to investigate the UFO phenomenon. As events unfold, comment will be added to this discussion, but it is suggested that people interested in UFO-related matters remain cautious and not let their emotions get the better of them until the outcome of these things can be appropriately analysed and critiqued.

But what about the theory that UFOs are extraterrestrial in origin? One of the compelling pieces of evidence for this is the UFO crash at Roswell. There are others, but this one will be talked about first. The military personnel involved in the recovery of crash debris say there were bodies in the wreckage. The bodies, they say, were unusual. What they meant by unusual was that the bodies were short, thin, and had large heads. That is certainly unusual, and it seems something highly trained individuals would be able to distinguish from 'normal' bodies, wouldn't you think? So why would highly trained personnel say there were bodies in the wreckage that didn't resemble 'normal' human bodies if there weren't? This is extremely puzzling, if not downright weird.

The implications of there being bodies in a UFO crash are far-reaching. There are two possibilities here. The first is that the bodies were human, and something had happened to them.

What that 'something' was is highly disturbing, to say the least, because the appearance of the bodies suggests they were distorted in some way. The shortness of the bodies is unusual and hard to account for. Perhaps they were children? Why would the heads be so exaggerated and the bodies so emaciated? This is very odd. What could have caused it? While the evidence is absent, the rumours are compelling. If the UFO crash is real, why not the presence of bodies?

The problem here is not the truth of a UFO crash or the presence of bodies. The problem is the *claim* that the bodies were extraterrestrial. This is the real problem here. Where is the evidence to support the claim? Once again, as with the UFO debris, it is under lock and key. No one has access to it, so no one can verify the truth or substantiate the claims. We only have circumstantial evidence supporting the claim that UFOs and extraterrestrials are real. Is this as far as we get along the evidence trail? Next, the discussion will focus on the experience of eyewitnesses.

6. Any evidence that extraterrestrials are real?

An incredibly famous and intriguing case of extraterrestrial visitation comes from Whitley Strieber and Betty and Barney Hill. Strieber is an author of fiction and has an interesting story to tell. He has authored books about his experience called *Communion* and *Transformation: The Breakthrough*. [Communion was made into a movie.] In Communion, Strieber tells us that he experienced strange events in his city apartment and on his property late at night sometime in 1985. He eventually realised (or believed) he was being visited by strange beings, who eventually kidnapped him from his bed and took him on a strange craft where they conducted weird medical experiments on him. Much of this was revealed through hypnosis and is, therefore, controversial.

Strieber tells us he had many such events until they settled down. Not sure what 'settled down' means but according to him they settled down because they 'stopped being sinister'. He seemed to have come to some kind of understanding with the strange beings; that everything would be fine, and he shouldn't worry that he'd seen something unusual if not highly strange. Or that he had experienced weird medical experiments. Though he suffered for many years with unusual psychology before coming to terms with it (in his own way) some years later, he sat down to write about his experience.

His experience was profound. He thought it meant something, or that he was compelled to feel that it meant something. He eventually thought he'd figured it out and wrote about it in *Transformation: The Breakthrough*. He believed the beings were 'spiritual' and had only wanted to 'commune' with him. Not just him, but with everyone on planet Earth. He now believed he understood them and what they were doing here and what they wanted. They wanted to commune.

Another intriguing case was that of Betty and Barney Hill [an upcoming documentary produced by Bryce Zabel (2022) which hasn't at the time of writing been released] – also being made into a movie. Driving home late one night from a weekend in Montreal in 1961, Betty Hill believed she was seeing a UFO; she called it a strange light. As they drove Route 3 through White Mountain National Park in New Hampshire, the strange light seemed to be following them and grew bigger the further south they went. At first, Barney thought it was a passenger plane bound for Montreal, and Betty thought it turned into a meteor.

However, as they headed for Franconia Notch, the strange light appeared to descend toward them, causing Barney to stop the car. They alighted from the car, and using the binoculars they had, saw that the craft was the size of a passenger plane. Barney claimed he could see strange-looking humanoids in the broad window of the craft. He counted perhaps ten humanoids standing at the window looking at him and Betty.

Barney claims most of the humanoids disappeared, leaving one still in the window. Barney then claims he thought he heard that humanoid talking to him, telling him to stay where he was and to keep looking. Aside from it being a strange request, the whole event is weird if not puzzling. It gets even stranger when some of those humanoids he'd seen in the window of the craft were suddenly walking toward him on the ground.

Barney panicked. He believed the humanoids were intent on kidnapping him and Betty, and he yelled at Betty to get back in the car before the humanoids captured them. As they sped away from the strange event, they realised one other completely bizarre thing had occurred, and this one thing has caused many to speculate that Barney and Betty Hill were abducted. They say that two-hours had transpired while they stood in the middle of the road observing the strange craft and the humanoids. Some say they 'lost' two hours, and this has become speculative meat for a phenomenon known as 'alien' abduction – lost time is claimed to be a symptom of being abducted.

This book will not discuss alleged 'alien' abduction because it is a side issue to UFOs and visiting extraterrestrials. A spin-off, if you will, of the whole UFO and extraterrestrial thing. Some see sinister written all over UFOs and extraterrestrials, and it has become a cottage industry fuelling movie-making and fanning the fire of alien invasion stories. But there are others, like Dr. Steven Greer and *The Disclosure Project*, who think all extraterrestrials are benign; they do not present a threat to humans, they come in peace. A feeling that eventually evolved out of Whitley Strieber's 'alien' visitations. But whatever is true of these strange and bewildering encounters, the 'experiencers' (a term coined by John E. Mack, the author of *Passport to the Cosmos*) all believe they saw something; something real. Make of that what you will. It has been touched on above, but where the discussion goes from here has nothing to do with 'alien' abduction stories or benign visits at night in the small hours by strange-looking creatures.

Why is this the end of the trail, so to speak? It is the end of the 'evidence' trail because no-one has an extraterrestrial friend they can call on to speak with others about their reality and existence. Or stand in the court of public opinion and verify anything. Neither does anyone possess a UFO or extraterrestrial craft for others to see and analyse, and this situation – reached years ago – has resulted in the development of another cultural industry called *speculative* Ufology. Speculative Ufology, if you didn't guess, is about speculating on UFOs and the existence of extraterrestrials.

7. Truth, exaggeration, misidentification, and lies?

By far the most challenging, even the hardest aspect of discussing UFOs and extraterrestrials, is figuring out what is exaggeration, misidentification, and lies in any investigation of UFOs, specifically, speculative Ufology. The misidentification aspect of investigating UFOs has been wonderfully covered by the work of MUFON field investigators. MUFON has shown that nine-out-of-ten UFO sightings are a misidentification, and that is significant. It is significant because it shows people are not seeing what they think they are seeing (nine-out-of-ten times). That is both good and bad for Ufology. It is good for Ufology because it removes unnecessary 'noise' from its object of study. It is bad because it leaves little doubt that UFOs are not what many people think they are, and that leaves MUFON with little to investigate.

Of the small number of cases that are not misidentifications, no conclusion can be drawn as to what was seen, and this has nothing to do with what was seen, but with the nature of the case. The eyewitness did not supply enough information for a proper investigation, and this is significant, because it hints at something to do with the person rather than what was seen. More about this in a moment. A MUFON field investigator, therefore, has their work cut out for them, and that means special education and training is needed. Like most professional fields of study, the investigator must be qualified.

But even with some education and training, one would think that, by now, at least, solid information on UFOs has been gathered. Not so. And that is unfortunate. What has happened since MUFON started collecting data on sightings, and developing an extraordinary case management system, is the collection of one type of data: sightings. It has collected to date 80,000-plus cases. That is impressive. But why hasn't this resulted in something substantial about UFOs and extraterrestrials? Why hasn't one of these sightings resulted in confirmation of MUFON's underlying theoretical premise and reason for existing: That UFOs are extraterrestrial in origin?

There is a pattern in the data collected, true. But the pattern does not lead anywhere specific. Well, nowhere for anyone with a strong belief that UFOs are extraterrestrial in origin. If one takes a scientific approach to studying UFOs, there are two patterns in the 'noise':

1) that people see things and keep seeing things;

2) that nine-out-of-ten sightings are a misidentification.

The sighting that isn't a confirmed misidentification cannot be 'identified' as a misidentification due to a lack of information. Therefore, the only conclusion that can be drawn from this data is that most people misidentify things that fly in the sky.

What has developed from this observation is grist for speculative Ufology. For instance, one MUFON field investigator (at the time of writing), Cheryl Costa, undertook a statistical analysis of the available data. This is good, for what, I'm not sure. It does, however, show patterns in the data. Cheryl has generated statistical data from the MUFON CMS to show numbers, places, times, and other nominals. For example, MUFON can now use statistics to show the number of sightings for a particular county (in the USA), down to the day. But how to read the statistics?

I'll examine a set of stats to figure how to read and use it.

Say there were 2,500 UFO sightings reported in May 2017 in Arkansas. So what? Well, perhaps Arkansas is the state with the highest reported incidents of sightings. This might suggest there is something special about Arkansas. But what? Do UFOs like Arkansas more than say, California? Well, this is one way to read the stats, but it would be an odd reading. For one, it leaves us with a bunch of odd questions to ask. For instance, why do UFOs like Arkansas more than other states? The problem with this is the word 'like'; it presumes to know what UFOs like, and that is a bit presumptuous. It's as odd as the rumour that spread a few years back that extraterrestrials love strawberry ice-cream. The presumptions are as obvious as the sun on a cloudless day.

What might be more significant, perhaps not for Ufology as much as for social psychology, anthropology, and sociology, is the gender demographic. Do more females report sightings than males in Arkansas? Or did more females than males report in May of 2017? There would be something to investigate here if there was a gender imbalance. Unlike the disapproving biologist Susan Freud (Facebook discussion, 2017) who scorned me for questioning everything she said, and, had the temerity to claim, even though she claimed to be a scientist, that, in the end, science did not produce facts. What a strange bit of dissociation. Of course, stats are momentary facts; they are facts the moment they were produced but not after that because they become 'trends' in the data – the numbers can instantly change, like the weather. So what was true a day ago may not be true the following day, or a week after that, and different (even false) a year later.

Repeat surveys are important for this reason, to reduce the margin of error or increase the confidence interval. (Statistics will be discussed but not in detail or in depth because stats are not the point of this book but may be included in the next edition or version of this book.)

Add to this statistical account of the number of sightings, the fact that nine-out-of-ten of those sightings will be misidentifications, this is the pattern produced by the investigation of UFO sightings. So of the hypothetical 2,500 sightings for Arkansas in May 2017, roughly 250 will be classified uninvestigable. Uninvestigable because of a lack of information. The rest will be shown to be misidentifications. So, what's the point? What's the point of MUFON if all it does is demonstrate that most sightings are misidentifications, regardless of the number, and the rest unknowns because they could not be properly investigated?

If, after 80,000-plus cases of sightings, when investigated, none produced solid data to conclude extraterrestrials were flying the friendly skies over Arkansas, what can we conclude? We can conclude that there is no 'scientific' evidence to prove the theory that UFO sightings are extraterrestrial in origin. That is the conclusion that can be confidently drawn from this simple discussion, and a dilemma for MUFON and others who claim to study UFOs. Will another 80,000 plus cases produce the 'scientific' evidence it seeks? Unlikely. Unlikely because the pattern has been set. The rest is pure speculation, and the speculation feeds into speculative Ufology and leads to going in circles.

Circulus in Probando.

As for Susan Freud's contention – the biologist (above) [Facebook discussion, 2017] – that extraterrestrials are visiting Earth and abducting people, spending the evening with them and watching TV or drinking tea, but without solid 'scientific' evidence, what can I say about her belief system? [Susan Freud alleges she has evidence but never furnishes it and pre-empts the discussion of the nature of 'scientific' evidence.] This is an important thing that is now touched on because this may be the most significant aspect of speculative Ufology, more than the rumours. That the people who claim privileged knowledge on the issue are a special type of person. They, of course, refer to their experience as the key to understanding the truth about UFOs and extraterrestrials. Cheryl Costa does a similar thing when she proposes a theory about the origin of different shaped UFOs.

Cheryl Costa (2017) claims that the shape of a UFO reveals its origin. She claims that by knowing the shape of a UFO, she can determine the origin of the extraterrestrials who made the craft and fly it. How is this possible? It's possible because she believes it's true. That's how. But how does she know that the shape of a UFO reveals its extraterrestrial origin? I don't know how she knows. There is a small question of 'scientific' evidence. Missing evidence, in fact. Take, for example, an elliptical UFO. The shape of an elliptical UFO means it is from the Pleiades. How can she know the craft is from the Pleiades? What scientific evidence does she have to prove that what she says is true? None. She has no scientific evidence. What she is doing is speculating, and this makes her a member of *speculative* Ufology.

What is wrong with this theory? It is a theory, and theory

will be discussed later. Well, first of all, there is no 'scientific evidence' to support her claim. Secondly, she assumes the UFO with the elliptical shape is from the Pleiades. This is audacious and speculative. Why would someone do this? Well, just like the naïve 'biologist' Susan Freud, who also behaves audaciously and speculatively when claiming extraterrestrials come from many places, including the Pleiades. Anything is possible ... in speculative Ufology. Susan claims to be a scientist yet ignores science and the scientific method to make outlandish claims about these things which are based on privileged knowledge (allegedly) that only she knows (and doesn't want to share with the rest of us).

Why doesn't Susan Freud subject her presumptions to scientific analysis? Why doesn't she create scientific experiments to test her theories? Perhaps because belief is more important than fact in speculative Ufology and when someone challenges her assumptions, she cries foul and invokes *begging the question* – a fallacy. The conclusion is embedded in her argument: if aliens didn't abduct me, who did? She is assuming aliens abducted her and her friends, and only wants to believe what she claims is true without subjecting it to science and the scientific method. This puts her in an awkward position, scientifically. To claim to use the scientific method yet not use it when it counts most, when it should be used, is a clear case of dissociation. She wants to believe but she doesn't want to prove.

What is science without systematic analysis?

Susan Freud claims to be a scientist, but is she? To claim to be a scientist means to search for answers to life's mysteries and solve life's problems (or create problems). What is science? Here's a definition for the non-scientist. According to Dictionary.com, it's the study of the physical and natural world using theoretical models and data from experiments or observation.

I'll start with the notion of 'theoretical models' because it isn't what it seems. I'll use Cheryl Costa's theory as an example and hope she doesn't get upset. Her theory says the shape of a UFO determines its extraterrestrial origin. There are two parts to this theory: firstly, the shape of a UFO; secondly, the origin of a UFO. To test this theory's internal credibility, we must show the factual basis of the theory's nominals. Firstly, are UFOs real? Yes, they are. Secondly, do UFOs come in different shapes? Apparently. Cheryl has produced a chart of UFO shapes, and has clarified this with *UFO Shapes, Documented* (Online). These are based on eyewitness accounts, not hers personally. Nothing wrong with that. Different people have seen these shapes. Thirdly, the shape determines the origin of the UFO. This is where her theory breaks down. There is no scientific evidence to show, as was demonstrated above (hypothetically), that an elliptical UFO originates from the Pleiades. This is an unsubstantiated claim. This means there is no scientific evidence to support the claim that evidence exists to prove the origin of a particular UFO by referring to its shape.

There is a missing link – the link between UFOs and extraterrestrials. This theory has not been substantiated.

This is the problem with baseless theories, and by baseless, I mean lacking scientific evidence, of which there are plenty in speculative Ufology. There are so many theories in speculative Ufology that are baseless, it isn't funny. Yet, the person constructing the theory believes the theory has a scientific basis. What is the scientific evidence that elliptical UFOs are from the Pleiades? There is none. Though Cheryl may claim elliptical UFOs are from the Pleiades, no-one, not even she, can substantiate the claim, and this is a shared problem in speculative Ufology. It claims to be scientific but doesn't use science and the scientific method to test any of its theories about UFOs and extraterrestrials, but its supporters will cry bloody murder if you challenge their assumptions.

Argumentum ad ignorantiam

What if I claim Cheryl plagiarised my book? Not only have I cast aspersions on her good character (which she does have, by the way), but I have claimed she has stolen my story and breached copyright laws. In a court of law, I would be required to produce the evidence of my claim. If I want to see justice done, that is. But that is a 'criminal' court. What court exists to defend allegations of scientific impropriety? The court of scientific peers (which may be considered biased) and the court of public opinion (which may be considered ignorant) – unless a patent is involved – that deals with issues of intellectual property, and perhaps plagiarism. These are the only courts that exist to pass judgement on claims of scientific impropriety. How easy is it to bring claims before peers and the people? Not that easy, and highly risky. Not only is your theory at risk, but so is your credibility and honesty. Cheryl risked a lot when she claimed the shape of a UFO determines its (extraterrestrial) origin.

What she is really doing is arguing from ignorance. As is everyone who claims UFOs are extraterrestrial craft. The person who makes such a claim also needs to supply the scientific evidence, otherwise they are arguing from ignorance. It is a fallacy to argue that because no-one can prove the claim is false, therefore (by divine right?) the claim is automatically true. It's a default position, and an unsound one; scientifically unsound, that is. Cheryl does not have an extraterrestrial friend who can verify her claims. Neither does she have a UFO that represents at least one shape she can use to back up her claims, and this is telling. It is also significant. It is significant because it underscores all the claims about UFOs being extraterrestrial in origin, regardless of their shape. If naïve biologist Susan Freud cannot produce, in her defence, an extraterrestrial 'friend' who can verify her claims about abduction, then she commits scientific impropriety. If Cheryl cannot produce a UFO to support her claim that it is extraterrestrial, regardless of its shape, she commits scientific impropriety.

These are mild criticisms, by the way. No one is going to take them to court over their claims.

Scientific impropriety, however, is wrong. It is wrong because it can deceive people and lead them astray, and leading people astray could be criminal, in the extreme.

8. Types of people in speculative Ufology

There are six types of people in speculative Ufology (that are credible): The liar, the hoaxer, the believer, the self-deluded, the naïve, and the ignorant. I place myself in the category of ignorant because I am ignorant of a great many things. I do not have all the facts about everything. Least of all, whether extraterrestrials are visiting planet Earth. There is no evidence to prove UFOs are extraterrestrial in origin, yet. Therefore, I cannot scientifically claim they are extraterrestrial. Neither can I claim they are not. Given that no-one else can prove they are extraterrestrial, where does that leave Ufology, if not speculative Ufology?

If Ufology is the scientific study of UFOs, then it uses science and the scientific method to study its subject: UFOs. If it claims to be scientific, then it uses science and scientific methods. However, there is little evidence that it does. Cheryl Costa's statistical analysis may be the exception. If you believe UFOs are extraterrestrial in origin, how do you investigate? Well, first of all, you must demonstrate that UFOs are physical objects and not imaginary ones. This, as was discussed above, has been done. UFOs are real objects. They are physical objects. [This includes anything that falls under the recent rubric of calling UFOs *UAPs*.] As to whether they are 'natural' objects, this also has been verified. UFO sightings, according to MUFON, are often misidentifications of 'natural' phenomena: planets, stars, meteorites, asteroids, and atmospheric phenomenon. In fact, nine-times-out-of-ten UFOs are 'natural' objects.

But are the one-in-ten manufactured objects? Hard to know, since the report of the sighting is incomplete.

Which UFOs are extraterrestrial craft? Unknown. No-one can say with any authority. There are only claims they are extraterrestrial, but no proof they are. When I say no proof, I mean no scientific proof, and this is significant. There is no scientific proof that UFOs are extraterrestrial craft. There is scientific 'proof' that UFOs are real, that they are physical 'natural' objects. Sometimes they are manufactured objects: planes, helicopters, satellites, drones, etc. This is not in dispute. What is disputed is the claim that UFOs are extraterrestrial craft, despite it being a 'logical' conclusion for some. The discussion is, after all, focused on the scientific study of UFOs/UAPs and not on other types of study. Scientific methods have been used to verify that UFOs are physical objects, but scientific methods are not used to verify that UFOs are extraterrestrial craft, which is the main point of this discussion. But, given the recent actions of the US Congress, this may very well change.

The liar

The liar. Liars lie. That's what they do. They tell lies. It is either pathological, in that they cannot help themselves, or it is deliberate – a distinction is drawn between pathological lying and deliberate lying. But when someone tells a lie about seeing a UFO, the lie is greater in its heinousness, because it is intended to deceive. It deliberately deceives. That is its aim, and it is criminal. It is criminal because people can be hurt by such lies, and not merely psychologically, like a lover deceived by a cheater. I'm thinking of cults when I think of this type of deception. When groups of people are led to believe that something bad is going to happen to planet Earth and their only escape is through suicide. Doomsday cults are famous (or should I say infamous?) for this kind of thing.

To lie about seeing a UFO is greater in magnitude because it has the potential to deceive large numbers of people if not an entire country. I'm thinking of H. G. Wells's hoax, *War of the Worlds*. People were hurt because they believed the radio broadcast and tried to escape. Not only did they hurt themselves, but they hurt others in the process.

The hoaxer

The hoaxer is like the Joker in Batman. A prankster. He goes out of his way to play a practical joke on naïve and ignorant people. The hoaxer sometimes doesn't see the harm in what he does, even if it leads people astray, while other hoaxers are more dangerous. Take, for example, the people who claimed to have a dead extraterrestrial in their possession and would reveal all in a conference – the Roswell artefact incident and see Pettit (2017). The organisers claimed to bring Truth (or should I say truth with a small *t*?). They hired a convention centre and charged people money for the privilege of seeing the evidence for themselves. Substantial amounts of money changed hands to put on a glamorous show.

The problem came when some people questioned the evidence. A chorus of disbelief ensued, an investigation followed, and the Truth (with a capital *T*) was revealed. The evidence had been faked. People had been deceived. People lost money. The hoax was criminal, and why the authors deserve to be punished. Criminal charges have yet to be laid, and perhaps never will. The organisers believed they'd been told the Truth (with a capital T). They claimed they were deceived. A convenient lie? Perhaps.

The believer

The believer is a different kettle of fish. The believer is someone who believes something is true without seeing the evidence. Most people in the UFO scene believe UFOs are real without having seen one. Most believers tend to believe UFOs are extraterrestrial in origin without having seen the evidence. This is called self-deception. They have deceived themselves. Just like a religious person believes in a divine being or God. A creator. They believe without evidence that such a being exists, that such a being did everything and does everything claimed of him or her. (Interesting side note: it is only recently that such entities have become gender-neutral.)

Self-deception is a cruel mistress. She is a cruel mistress because when the believer's belief is challenged, the believer will invoke a fallacy called *argumentum ad ignorantiam* to justify their belief. Because they engage in self-deception, they are unaware of the fact that they are ignorant in the belief. They do not possess evidence to support their belief but will argue with anyone who challenges the foundation of their belief. They will counter-claim that the challenger's claim is baseless because the challenger cannot prove the believer's position is untrue. They merely hurt themselves (or not) by maintaining the deception.

The self-deluded

The self-deluded are like believers. The self-deluded are people who simply believe what they want to believe, even if there isn't any evidence for the belief they hold. They are casually indifferent to demands for evidence. They care less about evidence. They simply hold onto the hope that what they believe is true.

These people are sometimes called post-truthers. A post-truther doesn't need evidence to maintain a belief in something. For example, they accept the stories of UFOs and claims of extraterrestrial origin without question. That extraterrestrials have been visiting planet Earth forever, they say. That UFOs are the craft they travel here across space and time or 'space-time' as it is now referred to, or across dimensions to come here. Some go further and claim that the UFO is a biological entity.

The naïve

The naïve are like the believers and the self-deluded. Only different. The naïve believe that if someone said UFOs exist, that's good enough for them. That if someone said UFOs are extraterrestrial in origin, that's also good enough for them. They go further, though, by saying that extraterrestrials are visiting Earth for a reason. That reason is to protect Earth from all the 'bad' people. That ETs bring warnings of destruction if Earthlings do not change their ways. Sound familiar? The naïve are closer to Doomsday cultists than either the believers or the self-deluded. For example, the Christian God will send you to Hell if you do not confess your sins and accept His Son Jesus as saviour and lord.

The naïve are proactive in getting the message out that higher beings are here to save us from ourselves. (Note: what does 'higher' mean?) That these higher beings want to commune with us but won't until we change our ways and start doing what they want us to do. The higher beings are always just out of reach, of course, or can be reached through meditation. The people who believe in the existence of these higher beings push the meditative arts and the like. If you want to get to a higher plane, the one the extraterrestrials are on – the naïve sometimes call them beings of light – then start meditating!

The ignorant

The ignorant are all of us. We are all ignorant of the truth (and Truth) of something. We are all ignorant of the truth (the Truth) about UFOs and extraterrestrials. And until we recognise this situation, we will not get close to the Truth. Not even a piñata. We humans do not know what UFOs are until we investigate them; investigate sightings and what are claimed of them. This can only happen when people do research using tried and tested methods of investigation. These are the only methods that will reveal the Truth about UFOs, in my opinion. MUFON, for example, does this (or tries to) – it investigates. MUFON's field investigators can show us the true nature of UFOs. And has done so, to considerable extent.

Note: a distinction has been made between truth and Truth. Truth, with a small 't', is personal truth. Truth with a capital 'T' is a fact, even a scientific fact. For example, a person believes UFOs are extraterrestrial in origin, therefore, their personal 'truth' is not a fact or even a scientific fact and therefore not Truth. Yet. It is quite possible for a person to hold many truths without knowing they are not scientific Truths, and vice versa. The above example could change in an instant if the people who know the Truth develop a conscience and reveal the Truth.

MUFON's system of data collection and field investigation is uniquely situated to establish facts. For instance, the fact that most UFOs are misidentifications. That's a fact (not personal truth). MUFON has shown through good investigative work that nine-out-of-ten sightings are misidentifications of naturally occurring phenomena, or something mundane. This is important. It's important because it provides a solid foundation from which to discuss the nature of UFOs and can be falsified, meaning replicated. And replication of findings is the basis of scientific research – all scientific findings must be replicated or demonstrated to be accurate by another researcher or researchers.

9. The science of studying the shit out of anything

"Science is a systematic enterprise that builds and organises knowledge in the form of testable explanations and predictions about the universe."

When I apply this idea to UFOs and extraterrestrials, I can at once see – and I hope others can – how whatever has gone on in the field of UFO research has failed to achieve anything like the above definition. As with anything, caveats apply, of course. Especially when it comes to testable explanations and predictions. Few individuals have done anything like this in this field, while some organisations have at least tried – for example, MUFON. MUFON is the preeminent group to systematically build and organise knowledge associated with its field of study. There may be others. Well, MUFON doesn't actually build it; it collects and organises it. On the other hand, it does organise knowledge of its subject. It has a data collection system that collects and collates reports of UFOs. MUFON calls it the case management system, or CMS for short.

MUFON's CMS contains roughly – at the time of writing – 80,000-plus cases. This is extraordinary and significant, in two main ways: 1) people keep seeing things (and reporting the sighting); and 2) if people continue seeing things, then it begs to be investigated. Few organizations, including research institutes, would boast such a database of cases in any scientific field. Yet MUFON and others suffer from a nasty malady: scientific credibility. This means, that while the CMS has roughly 80,000-plus cases, little to no research on the cases has yielded testable explanations and predictions about them.

Well, except one: people keep seeing things. This is significant, of course.

While a recent effort has been made to generate statistical data about the information in the CMS, it hasn't resulted in anything more significant. For instance, while it is interesting that the statistics show high sightings – allegedly increasing numbers, according to Dunne (2017) – in certain locations (e.g., the USA, mostly), that is all that has been achieved. What probably can be achieved. If one were to generate a testable explanation for the number of sightings, however, it would merely be that nine-out-of-ten cases can be explained as something mundane. That is a testable prediction.

And that is a problem for MUFON.

MUFON (and anyone who claims to study UFOs) has a problem, in that, if it were to make a prediction about the cases in its CMS, the prediction would be that the pattern of cases is benign. If nine-out-of-ten cases can be explained as something mundane and unrelated to its aim of studying the phenomenon it purportedly collects information on, then it is not studying UFOs. Nothing significant is generated by the data – except for the stats, perhaps. The most basic pattern in the data is that *people see things.*

Ordinary, mundane things, as it turns out.

It is quite likely, too, that if one were to test the explanation provided for such an enormous collection of cases, it would be that the next 80,000-plus cases will be the same. One could probably predict this accurately if one so chooses. And that is puzzling. It is puzzling because the biggest claim in the study of UFOs is that UFOs are extraterrestrial in origin. A claim based on the belief that humans couldn't possibly build such technology. If no data can be gleaned from 80,000-plus cases that link extraterrestrials to UFOs (other than the claim that extraterrestrials are visiting some people and abducting others – not trying to dimmish the significance of abductions, here), then MUFON cannot claim that there is a link, and is maybe confused about its purpose for existing.

In order to claim to be scientific, MUFON must use scientific methods to study its subject, UFOs. However, the methods of investigation are limited to investigating the *claims* someone has made about a sighting. There are other things to investigate than what someone claims about the thing they saw, by the way. And this is something many people misunderstand about the study of UFOs, I think. If one claims to study something, he/she must show the thing being studied. For instance, if he/she says they are studying 'cells' then they provide examples of the 'cells' they are studying – whether they be human cells or tree cells. The study of cells is called cytology. Likewise, if he/she says they are studying UFOs, they must show the UFOs they are studying.

We can immediately see the problem with any claim to *study* UFOs.

In addition, there is a missing link. And, incidentally, the most significant part of the study of UFOs. There is a missing link for the people who claim UFOs are extraterrestrial in origin. If UFOs are extraterrestrial in origin, then MUFON and those who claim to study UFOs as craft from elsewhere, must show that link. If they cannot, they cannot claim to be engaged in the scientific study of UFOs. MUFON cannot make such a claim because it is doing something that is already being done by other (bonafide) scientific groups, and I mean the study of aerial or atmospheric phenomena, e.g., meteorology, astronomy, and astrophysics. To study the same thing under a different name and claim it is scientific without using scientific methods is called misdirection, and, perhaps, for good reason.

Am I casting aspersions? You bet.

When MUFON is challenged (or its members challenged) to explain its aims, it simply says it studies UFOs. If you ask it what a UFO is, it invariably says an *unidentified flying object.* What is so significant about that? Well, because of the claim made about UFOs, that's what. MUFON claims they are extraterrestrial in origin – even if some of its members claim they are spiritual entities. If UFOs were not believed to be extraterrestrial in origin, then no one would bother studying them on the pretext that they were extraterrestrial. If this were true, it would be the greatest and most significant event in human history. And a damn good reason to investigate them. MUFON was founded in May 1969 to make the organisation appear more scientific in its outlook if not in practice. But there is a slight problem with this claim: what is its *object* of study?

What? I hear some readers ask. We know what MUFON's subject is, UFOs. But what is its object of study?

If one looks at the data generated by MUFON's CMS, it will become clearer that MUFON's object of study is *sightings* of unidentified flying objects (see MUFON's 'Mission Statement and Goals'). Because this is the thing that people claim to be seeing and report the sighting to MUFON (or another agency). But what is MUFON studying? MUFON investigates *sightings* of UFOs. It does not actually study the UFOs that were seen. And this is important because its focus is on what people *claim to see* and not on what anyone has in their possession, e.g., an actual UFO. MUFON doesn't have an actual UFO; no one does. Aside from the conspiracy theory that someone does, and that someone is someone in an unidentifiable government agency who won't show it to anyone. And hence, a conspiracy theory.

Nothing wrong with conspiracy theories in an age of global fascism/totalitarianism. Asking questions is normal and should not be viewed as doing something wrong. Questioning our leaders' motives and agendas is not a 'crime' against those with power and authority. That's all I say about that here.

There are many problems with this situation. If MUFON merely investigates what people claim to have seen, then what they are doing is investigating peoples' claims. They are investigating peoples' claims and not the thing they saw, e.g., a UFO. Investigating peoples' claims is not the same thing as investigating UFOs. If one investigates peoples' claims, it will be to verify the claim made by the person, that they saw something, was true. But this only begs the question: What is MUFON's aim? Is its aim to study UFOs? Or is it to study peoples' claims? If it is the former, then it doesn't do that. What MUFON does is investigate peoples' claims of seeing something, which it clearly does.

After collecting 80,000-plus cases, can MUFON claim to have actual scientific evidence that people see things? Yes, it can. What is the evidence? Eighty-thousand-plus sightings in its CMS (that have been investigated to the best of its ability). It also has investigative fieldwork (in the form of reports) that show that nine-out-of-ten sightings are of a mundane nature – explained as something mundane. Therefore, it has a lot of data about mundane things. But does it have any scientific evidence that what anyone saw was an extraterrestrial craft? No, it doesn't. Its investigators have neither seen the UFO witnessed nor an extraterrestrial connected with an abduction or visitation (if seen along with the reported UFO sighting). And that is significant.

Therefore, the claim to study UFOs is oxymoronic. To claim to scientifically study UFOs is confounding. Why? Because MUFON investigates sightings of UFOs and not actual UFOs. Thus, it is more like a private investigator or contractor that does police work. It is more like police work because this is what the police do, investigate crime. (Not all UFOs are crimes, by the way. But some are.) This is a crucial point because what UFOs do can be considered criminal. For one, they violate airspace, and that's a crime. If the pilots are abducting people, this too is criminal.

Firstly, UFOs enter a country's airspace without the permission of the authorities. Secondly, by breaching a country's airspace, they increase the risk of danger to air traffic if not ground traffic. But I'll stay on course because these issues are peripheral to the main story.

What is of concern is what is being investigated. For a group or organisation to claim to be scientifically studying UFOs, it must be studying the object of its focus. In this case, UFOs. However, as I have just argued, no one is studying UFOs. If they were studying UFOs, they would have at least one UFO in their possession (built by them or offered by the owner) and under the microscope, as they say, just like if they were studying the bones of dinosaurs. But no one has a UFO. If nine-out-of-ten UFOs are mundane things (e.g., misidentified of natural phenomena), then the study of UFOs is the study of mundane things. It is not the study of UFOs as extraterrestrial craft.

Of course, natural phenomena are not mundane things to naturalists or astronomers. But they are not the concern of those who claim to be studying UFOs.

To study UFOs, the studier must have one (or more) to study. If not, they are not studying UFOs. If he/she claims to be studying UFOs (even indirectly), what are they studying? If they say they are collecting cases of sightings, reviewing old documents, and the like (secondary data analysis or archival study), and studying those, then they are, in fact, studying people, because people have sightings and write documents. They are studying claims made by people about seeing something or being a witness to something. And the only conclusion that can be drawn from this is whether the witness did see something, and what they saw. Nine-out-of-ten sightings are of mundane things. Therefore, ninety- to ninety-five percent of the time they are studying peoples' sightings of mundane things. Nine-out-of-ten sightings can be explained as something mundane, therefore, the chances of them studying something else are small to zero.

We can predict with great accuracy that most sightings will be of a mundane nature. Given this outcome, what is left to study? If you say the small percentage of sightings that aren't explainable (2 to 5%; some say 20%), how are you going to study the unexplainable? What I mean is, if nine-out-of-ten sightings are explainable based on significant data and experienced investigative work, then the small percentage of unexplainables will be much harder if not near-on impossible to verify, simply because there isn't enough data. And remember, the study of UFOs, at present, is the study of sightings or peoples' claims about what they saw. Therefore, what all investigators are really doing is verifying whether the sighting (or document) was genuine or not. To do that, the investigator must study carefully what the person says about the sighting, or what was said in the document.

If you say the sighting was verified by a neighbour or other witness, and radar, these are important. They are important because they substantiate what the witness said, to some degree. And confirm they saw something. But that's as far as the investigation will ever get because the seen object isn't available for further study – except, perhaps, says Leslie Kean (2011), when you have multiple sightings of the same thing. Without the object, though, the UFO, the investigator is hamstrung in his or her investigation, and the breadcrumbs stop. It always ends without the object ever being found. And that is a huge problem for anyone claiming to study UFOs. And, like a police officer investigating a crime, without the culprit, the case cannot be closed.

And so it is the same with all investigations of sightings of UFOs. All cases must either be closed or forever remain open. If MUFON claims to have closed a single UFO case, it can only be because the object or the UFO, or the mundanity, was found. Nine-out-of-ten cases have been closed because the object, or the thing seen, was found. The one-in-ten cases that stay open are open because the object, the seen thing, was never found. Without a UFO, however, all cases must remain open until the object is found.

What are the chances that the one-in-ten case is a UFO of

extraterrestrial origin? Hard to know, really, without the UFO and its pilot. And that is the mystery that may or may never be solved. Not by human beings, at least.

10. Methods and methodologies

One of the useful things MUFON did was create the case management system. Why? Because it allowed MUFON to collect more data about UFO sightings than would otherwise have been possible. This method saved a lot of hard footwork – investigators responding to rumours and hearsay and the occasional police report. That was old school. Nothing wrong with old school, but it would not have resulted in a database of 80,000-plus cases in the time between the earliest deposited sightings, around May 1969, till now. Some four-and-a-half decades of data collection.

The CMS is a valuable tool in the study of sightings of UFOs, if not actual UFOs. Its value is therefore limited. It can only collect data about sightings, and that's all. As Cheryl Costa, a member of the MUFON team, has demonstrated, useful statistics can now be generated from the CMS – case management system. And that is both significant and important. It is significant because it tells us numbers, the number of sightings on any day in any place. And it is important because the data can then be cross-referenced with other data. For instance, planet and star locations, asteroid locations, meteor showers, general and specific air traffic, and weather balloons.

Correlating sighting data with other data is especially important. This is because an investigator can systematically match sighting data with the mundane and rule out anything innocuous. Ruling out the innocuous is important. And nine-times-out-of-ten, this is what MUFON field investigators do. Ruling out (nine-times-out-of-ten) the innocuous demonstrates that people are prone to misidentify the things they see. Two or three professional investigations conducted by the government and military concluded that UFOs are innocuous. (It has been said that these 'official' investigations had an agenda, what that agenda was, has yet to be clarified.)

The point I'm making here is that the study of UFOs is really the study of peoples' sightings, which is akin to police work.

For those who don't know what research methods and methodologies are, here is a simple explanation. A method is the tool used to answer a research question (e.g., Are UFOs real?). A methodology is the strategy and rationale for using the method chosen to answer the research question [no justification needed for asking the question because all questions are up for grabs, answering (or not the question) will provide the justification for asking it] (e.g., Are UFOs real? This question will be answered with method A – observation/participant observation; survey; interviews; focus groups; experiment(s); secondary data analysis/archival study; a mixed methods approach). The methodology discusses why the research question should be answered (because it will shine a light on X), and why method A is the best way to answer the question, and whether qualitative or quantitative data will be collected – qualitative is usually 'themes' or 'categories' found in verbal or written texts. For example, an interview of twenty-five people produced three different 'themes' and were placed in category A, B, or C, or the text revealed something about Z. The methodology also includes the limitations of the research and the reliability of the method used (e.g., bias [a white guy interviewing black guys], the validity of the method [how valid is the method for collecting the data], the chance of failure/success [the likelihood of getting enough data]) and other interference (e.g., factors affecting obtaining the data to answer the question – the number of interviewees, the number of surveys, the type of document to be analysed, the weather).What other methods and methodologies are used to study UFOs?

There is a method used to initiate contact with the pilots of UFOs, called CE-5 – a close encounter of the fifth kind. The category was invented (as far as I know) by Dr. J. Allen Hynek (2001). The first time the idea was used (that I'm aware) was by Steven Spielberg for his movie, *Close Encounters of the Third Kind* (1977). It concerns close contact with the pilots of a UFO. Initially, because the pilots of the craft initiate contact, not the person – and usually unsolicited (e.g., buzz aircraft, buzz cars, hover above houses and farmland, or buzz airports). When people initiate contact, however, and contact is made with an extraterrestrial (allegedly), it's called a close encounter of the fifth kind or CE-5.

The best-known unsolicited encounter with extraterrestrials was that of Whitley Strieber. There are others (e.g., Travis Walton; Betty and Barney Hill), but Strieber's is the most celebrated. Strieber wrote about his experience in two books: *Communion* and *Transformation: The Breakthrough*. The first book details his experience and was turned into a movie starring Christopher Walken. This is a first-hand (and personal) account of a CE-3.

Dr. Stephen Greer, a medical doctor by training, has tried to make it a reality (through experimentation and immersion [participant observation]), for anyone wanting to experience it. This approach underlines his belief that all extraterrestrials are benign. A problematic view, to be sure – it ignores evolutionary processes and ethics. He has become a prominent figure in the field of UFOs, firstly, for his Disclosure Project (2010). The Disclosure Project aims to raise public awareness and pressure governments (through the media and FOIA) to release withheld documents and 'evidence' pertaining to UFOs and the (claimed) extraterrestrial presence on Earth. He collects eyewitnesses with 'credibility' – people who would be considered reliable and trustworthy by the general public – and interviews them and markets their experience through videos like *Sirius* and *Unacknowledged*. A recent documentary called *The Phenomenon* is another but made by James Fox. Dr. Greer has amassed an impressive range of people.

These people have mostly had a CE-3 – a close encounter with a UFO.

The method Dr. Greer uses to gather data on the subject is called CE-5, and uses two aspects: 1) experimentation, and 2) immersion. The experimental aspect involves using an app with 'sounds' gathered at a particular UFO event. The sounds are played and, the theory is, the pilots of UFOs come to the sounds. The 'immersion' aspect is that people immerse themselves in a CE-4/5 event through meditation – they either see a UFO or see an extraterrestrial or both. However, Dr. Greer's work is highly controversial. But not to the people who experience it. CE-5, therefore, refers to a 'fifth' level of UFO experience – human-initiated contact with the pilots of UFOs.

There is one other person who should be mentioned before we move on. This person is also a controversial figure in the history of the study of UFOs. Mostly because of what he seems to have achieved. The person is Stan Deyo (1996). He authored a book called *The Cosmic Conspiracy*. It's the ultimate conspiracy theory book to have ever been written. It's based on his 1969 near-death experience in which he has a vision of the 'future' dominated by an 'evil' alien presence on Earth.

Among his many claims is the one about having discovered an electrogravitic propulsion system. And he claims it is the one the 'evil' aliens use in their spacecraft or UFOs. His book supplies details of the propulsion system as well as the math to back it up (allegedly). The most controversial aspect of his story is the religious aspect. He claims that the UFO scene is nothing more than the work of Satan and his minions and will be used by a few sinister groups (e.g., the Economic Forum) to take control of humanity at some unspecified time in the future. That future time is claimed to be the one spoken of in the Bible, called the End Times, or the Rapture by fundamentalist Christians.

Now aside from Deyo's theatrics and religious conspiracy theory about evil entities coming to take over the world, he claims to have proven his theory about electrogravitic propulsion by building his own propulsion system used by UFOs. This is significant and could explain a great many UFO sightings. If Deyo has built (or did build, since he's now deceased) his own UFO, then he's probably been flying it about at night, scaring the socks off people. However, since the publication of his book *The Cosmic Conspiracy*, and according to Salla (2016), not a peep has been heard about it. And that is puzzling. (It is claimed that the 'authorities' (whoever they are) confiscated his work.

If Deyo had wanted to prove that his theory about the propulsion system was true, why didn't he show it to the world? Why didn't he make it available for inspection and analysis by interested and curious engineers, mathematicians, and others? He has focused, according to his website, not on his propulsion system, but on the religious conspiracy theory. According to the website, he has focused not on the conspiracy theory and other things and hasn't shown his research to anyone. This just makes him look like most people allegedly studying UFOs, all vim and vigour about UFOs but he hasn't shown the goods.

And (conspiracy theories aside) neither has anyone else. [Note: Since writing this, the US Navy has released some videos allegedly showing UFOs.]

11. Ufology is the study of people

I was privileged to be a Facebook friend of Rich Hoffman's, a veteran member and field investigator for MUFON. He has investigated many sightings and written up the reports. Rich is an engineer by profession and has a history and much experience in the field of engineering. He spent years in the field of 'Ufology' and gained insights into many a UFO sighting. But like most people, he has never seen a UFO up close and personal. He has never inspected one, never analysed one, never conducted experiments on one. Yet he continues to dedicate his time and energy to the phenomenon.

However, he recently ended his MUFON membership, as did many people. (I had renewed my membership earlier in the year, but let it lapse after the advent of a controversy.) The reason was due to (alleged) insidious 'racism' in the ranks of MUFON. Not only 'racism' but quite possibly fraud. MUFON's 'inner circle' may be intentionally thwarting the aims of the organisation. This last point is more damaging than the former, I think.

The organisation has structure. There are members and there are staff. There is also a mission statement. But few realised there was a murky centre to MUFON; it's called the 'inner circle'.

The 'inner circle' are people who've paid extravagant membership fees and been given privileged access to MUFON's leadership. (Hearsay?) And this has proven controversial in the extreme. If a person can afford to pay the extravagant membership fee, they also get a say in the direction MUFON takes. It says so in the fine print. Yet, when its illustrious (present – at the time of writing) leader was interviewed about it, he said it wasn't the case. But that wasn't all. He was also said to laugh off the 'controversy' as a joke, as being of no real consequence.

Members declining to renew or cancelling their membership in MUFON were surprised to learn this. They originally left to make a statement about the insidious 'racism' in its ranks. They were surprised to learn of the 'inner circle's' sway on leadership and input on MUFON's direction. It seems that many people didn't appreciate this revelation, given the 'inner circle's' characteristics. Some of the members of the 'inner circle' have questionable and worrisome backgrounds – scientifically speaking.

There are spiritualists, experiencers, and non-scientists occupying the 'inner circle'. If they have a say in MUFON's direction, and the ordinary members do not, it could cause a catastrophe for MUFON. One doesn't need to be a rocket scientist to figure out the direction the 'inner circle' will take MUFON. If it hasn't already. And there are troubling signs it has, and the catastrophe is a media headline waiting to be printed.

If MUFON's claims are contradictory, it cannot achieve anything worthwhile. It can't achieve anything remotely scientific. And that is a shame. Given the situation, MUFON needs to reassess its place in the world and its purpose for existing. Its credibility is slowly being eroded. And the erosion may have speeded up now that people aren't renewing their memberships. (Of course, this trend may reverse.) Its dirty little secret has been aired in public – the 'inner circle' will be its downfall. Unless, of course, MUFON recognises its leadership has failed to provide the kind of leadership offered in the glossy brochure: professional and scientific.

Given the study of UFOs (as I've argued) is really the study of peoples' claims, MUFON's 'inner circle' should be made up of professionals and scientists. There should be sociologists, psychologists, neuroscientists, and others in the 'inner circle', just to name a few. This is aside from the usual suspects: engineers, mathematicians, physicists, and the like. By not having these people in the 'inner circle', guiding it, MUFON has created a mini World Council of Churches – a flavour for every non-scientist.

MUFON's situation is telling, given the rise of the anti-science brigade in the USA (and elsewhere). Even the nation's President is anti-science (now ex-President, Donald Trump). MUFON's 'inner circle' is made up of anti-science people. And that reveals what is on the minds of MUFON's leadership: 1) the study of UFOs isn't (really) scientific; 2) UFOs aren't a scientific problem to solve; 3) UFOs belong to the 'spiritual' dimension; and 4) extraterrestrials aren't abducting people, lessening the suspicion around abductions.

The first is self-explanatory, to a degree. Given that the study of UFOs is the study of peoples' claims, why bother being covertly scientific? Secondly, if nothing scientific is going on behind closed doors, then no research is being done to solve the mystery of UFOs. Thirdly, if UFOs belong to the 'spiritual' realm, then MUFON leadership's judgement has been clouded. A clouded judgement lets all-manner-of-silliness take hold in its leadership and the organisation. Especially (baseless) conspiracy theories about the nature and reality of UFOs. MUFON's leadership is more likely to agree with the likes of Stan Deyo and Whitley Strieber in extremis than accredited scientists.

The fourth aspect is troubling. By letting 'experiencers' determine MUFON's direction, it lets an extremely controversial mindset direct its aims. The abduction of people is not only unethical but, as I've argued, is a crime. Of course, MUFON (earlier this year) seemed to anticipate this when it changed its policy (or planned to) on how it views 'alien' abduction. In early 2017, MUFON declared it would stop viewing 'alien' abduction as abduction. Instead, it was going to view it as an 'experience'. Hence the term 'experiencer'. The term conveniently sidesteps the 'ethical' and 'criminal' nature of the experience, in my humble opinion.

It also places the 'blame' on the 'experiencer' rather than the perpetrator of the experience. If the extraterrestrial is not the perpetrator, then the human is. If humans are the author of their 'alien' abduction experience, then it isn't the extraterrestrial. And it agrees with Randle, Estes, and Cone (2000), the authors of a book called *The Abduction Enigma: The truth behind the mass alien abductions of the late twentieth century*. This book claims to have (scientifically) investigated the 'alien' abduction phenomenon and concluded that it wasn't happening. That the abduction experience was happening in the heads of 'experiencers'. Their methodology is critiqued (not perfectly) by me (2012) on my WordPress website.

Although MUFON's 'inner circle' is made up of 'experiencers' and the like, the organisation insists it is involved in the 'scientific' study of UFOs. How can this be when its 'inner circle' consists of spiritualists and 'experiencers', and not a single accredited scientist on the horizon? MUFON's leadership is either complicit in its contradictory nature, or it is suffering some psychological malady. A malady called 'dissociation'? Perhaps.

More than anything, MUFON appears to be suffering from 'multiple personality disorder', MPD for short. MUFON is like the Vatican. On the one hand, it claims to be the voice and soul of Christianity on Earth – God's representative on Earth. It claims to adhere to the 'truths' of the Christian Bible, as dictated to its 'members' (its clergy, not ordinary folks) by God himself. Yet, behind closed doors, Willan (2017) says its members engage in 'wild gay orgies'. (Again, hearsay?) And before you get that look on your face that suggests I'm being anti-gay, au contraire; I'm a supporter of the LGBTQ community. I see nothing 'unnatural' about gays expressing themselves and indulging their natural desires, even if they are members of the Clergy and the Vatican. What I'm alluding to is the Vatican's 'dissociative' behaviour; its leadership suffers MPD.

12. How can Ufology get scientific?

It cannot, is the short answer. Well, maybe. Dr. J. Allen Hynek fans would disagree. The long answer is that if someone or a group can manage to get hold of a UFO (or two) and make it available to researchers, then Ufology stands a chance of fulfilling its stated aim: the study of UFOs. Until that day happens, anyone claiming to be scientifically studying UFOs can be dismissed as overly ambitious (if they don't use the scientific method and scientific methods of enquiry). Without a UFO, no one can study UFOs. End of story. The study of reports of sightings is not the study of UFOs; it's the study of people and peoples' claims.

It is probably best if MUFON and other people and groups claiming to be engaged in a scientific study of UFOs call themselves something else.

Likewise, if someone claims UFOs are extraterrestrial in origin, we can dismiss them as unscientific. That is not being rude, it is just clarifying what is and is not scientific study. This is not about personal belief, but about science. There is a missing link in the theory that UFOs are extraterrestrial in origin, so no one can claim UFOs are extraterrestrial in origin until it is found. And, likewise, for Ufology, someone or some group must find an extraterrestrial who is willing to 'disclose information' and clarify and confirm that UFOs are extraterrestrial in origin. Until this day comes, we can dismiss any claim that UFOs are extraterrestrial in origin on any grounds. This too is not being rude; it is distinguishing between fact and conjecture.

UFO: THE MAKING OF A MYTH?

How have UFOs come to be linked with extraterrestrials?

The most common way UFOs are linked with extraterrestrials is technology. Whenever anyone (including myself) has seen something that does not fit common notions of technology, they say it has got to be extraterrestrials because humans cannot possibly have developed such technology. UFOs are unconventional technology. This belief reveals a distinct ignorance about what is and what is not human technology. Just because it has not been seen before, does not mean it is not human technology.

Shades of Mary Shelley's *Frankenstein*? Perhaps.

Technology is changing so rapidly these days; it is almost impossible to keep up. I can't, and claim to be at the forefront of scientific development. I can barely keep up with my specialisation, of course. And I can't claim to keep up with any other field. Though I have an interest in certain fields of research, I work hard to remain informed. And that is challenging enough as it is without intentionally trying to keep up to date with everything.

Just look at how far technology has advanced since 1970, the year after humans landed on the Moon. (And NASA is attempting to do it again!) The advancements in computing have been phenomenal. And recently, Thompson (2017) says the Chinese managed to teleport a particle from Earth to space, a major breakthrough in the theory of teleportation. This kind of thing was a dream in 1970 – check the TV Show, *Star Trek* – but is now becoming commonplace. There has been advances in materials, and now we have nanocarbon-based materials that defy the materials used to make the rockets that sent men to the Moon. They are used in military planes, which do things that planes in 1970 could never achieve (e.g., clarify the F-35 Raptor).

One last one, that seems obvious when mentioned, is drone technology. Drone technology has evolved from the original unmanned or automated (non-piloted) craft that delivered pleasure to modelists (e.g., model planes) to aerial reconnaissance surveillance video to dropping bombs. Such technology has spread to cars and trucks. Where will it be used next? The last is the hovering drone that has evolved from the small quarto-prop model, minus a camera to with a camera, to be able to carry a person or more than one person. Many of these have been declared UFOs because the owner decorates the drone with coloured lights and flies it down some dark lane in the countryside. So to say UFOs must be extraterrestrial simply because the person who saw it can't imagine the technology is possible or even achievable today, is, quite frankly, a revelatory ignorance.

Another reason for saying UFOs are extraterrestrial in origin is the presence of extraterrestrials. The first extraterrestrials were, allegedly, those seen at the Roswell incident. Some were dead, had died in the crash, while others had survived and were alive. Unfortunately, we will never know because the military (allegedly) took the bodies and the living entities and hid them deep inside the military complex where the public cannot see them.

There is also the alleged contact and abduction by extraterrestrials. However, controversy surrounds what is really happening. MUFON, for instance, used to believe abductions were happening, but recently changed its mind. It has decided, instead, to call abductions 'experiences', borrowing the term from John E. Mack, M. D., and the abductee has come to be viewed as an 'experiencer'. The John E. Mack Institute also runs a program called PEER – Program for Extraordinary Experience Research. The point I'm making here is that if someone gets taken against their will, then it is an abduction, pure and simple. But to call such an experience merely an 'experience' is being overly politically correct, I would think. And even since the earliest accounts of visitation or abduction (the early 1950s?), the stories of abduction have multiplied. A recent but controversial story involves a man called Stan Romanek (2017).

I won't go deeply into Stan Romanek's story because I promised not to talk about extraterrestrials – not until the link between UFOs and extraterrestrials is substantiated. But Romanek claims to have had contact with 'aliens' since as early as the year 2000. He said extraterrestrials followed him home and personally visited him. He attributes 'mysterious' wounds on his body to extraterrestrials, and, more controversially, claims to have experienced 'telepathic' communication with extraterrestrials. Controversial, not because it cannot be scientifically shown as possible, but because this is something that happens in the person's mind, not external to it. John E. Mack would say Romanek was an 'experiencer'. But if, as Romanek claims, the extraterrestrials 'wounded' him (in many ways on many occasions), the encounter was not very ethical. Causing grievous bodily harm is against the law.

Why would MUFON distance itself from ethics?

If most people who've had personal contact with an extraterrestrial experienced it in unlawful ways, then extraterrestrials are not very ethical. See my article (2012) about this very issue. For MUFON to relegate such an experience to something the 'abductee' merely experiences is downright weird. To do so puts the onus of responsibility for the event squarely on the shoulders of the 'victim' and not the perpetrator, the extraterrestrial. Any unsolicited interference from an extraterrestrial is unethical and unlawful. Extraterrestrials who interfere with a person's life are unethical. And it shows complete disregard for the person's sovereignty. And this also applies to any alleged UFO of extraterrestrial origin that enters the airspace of a sovereign country.

Now the reader can see (I hope) why I won't discuss UFOs as extraterrestrial craft. Such behaviour is very human. I'll leave that thought right here.

If we discuss UFOs as extraterrestrial craft, we must deal with the issue of breaching sovereignty. And that is something no one in Ufology wants to do except me, that I'm aware of. I don't know, maybe there are people out there who have discussed this issue, but I haven't heard from them or seen their articles about it.

Although UFOs are a strange phenomenon that defies rational thinking, they are technology – advanced technology – and must be explained – an answer given for them. Either they are human technology, or they are extraterrestrial technology. They are not phantoms or will-o-the-wisp, despite what the 'spiritualists' and 'mysticists' may believe. If we accept this definition, we are not dealing with extraterrestrials. End of story. If they are human technology, however, then the designers and makers of such technology need to come clean about what they've done, because their tech isn't being used in an ethical way. If the military is using the tech to conduct weird experiments on people – deceiving people into thinking they are extraterrestrial and conducting weird medical experiments on them – then the military should apologise for the deception. If they don't, they are morally bankrupt. Perhaps they think they do so for the greater good? Perhaps they think their behaviour constitutes a means to achieve military superiority against an 'enemy'? And this justifies the psychological trauma? If not the physical trauma?

Therein lies the dilemma for Ufologists.

To accept UFOs as will-o-the-wisp, or to accept them as advanced technology. If Ufologists take the first path, then they cannot do anything scientific (at present), because such things belong to the realm of fantasy and imagination. They belong to the realm of the mind. This then is a subject for psychiatrists, psychologists, neurobiologists, and neuroscientists. Perhaps anthropologists and archaeologists? If, on the other hand, UFOs are technology, then they logically belong in the realm of other science, e.g., physics or engineering. MUFON believes UFOs should also be the concern of biologists, chemists, geologists, statisticians, computer scientists, and astronomers. It believes this because there is sometimes 'physical' evidence left at the scene of an alleged UFO sighting or an alleged landing.

All human technology comes from human imagination. Unless you're a conspiracy theorist and believe, as some do, that much new military technology (in the USA, at least) is the result of reverse engineering UFOs – the ones believed to be kept in some secret military base according to Robert Lazar (See 'Bob' Lazar – Robert Scott Lazar). This implies 'the US military' has a UFO, or more than one, in their possession.

While MUFON believes it studies UFOs, it engages in a

form of applied science. It doesn't study UFOs directly, it studies UFOs indirectly – using the term 'study' loosely – through the reports of UFO sightings. It therefore imagines what the technology must be like for the craft to do what it does. This links back to Stan Deyo, the only person known to (allegedly) understand the mathematics involved in the technology. The only person known to have solved the mystery of UFO technology and known to have made a UFO propulsion system, or 'flying saucer' engine, if you will. I know of no other person to have done so.

Hypothesising

Of the two hypotheses about UFOs, the first is easier to accept than the second. The first hypothesis says UFOs are human technology. I'll call it H1. A difficult hypothesis for many UFO-believers to accept. The second hypothesis says UFOs are extraterrestrial technology. I'll call it H2. A difficult hypothesis for many sceptics to accept. Mostly because it's easier to believe humans created the technology rather than extraterrestrials – regardless of how illogical that sounds. Why wouldn't extraterrestrials develop technology? H1 is harder for UFO-believers to accept because the technology takes on a 'magical' appearance rather than a 'rational' one. They don't want to accept that humans could possibly have invented the technology yet. But the scientific discoveries of the past thirty years or more make it easier to believe that humans could have invented the technology.

This doesn't mean that extraterrestrials didn't also invent the technology. It just means that it is the harder of the two hypotheses to accept without question. Firstly, to accept H2 is to accept that extraterrestrials are technological beings, not ethereal or spiritual beings. Many UFO-believers think extraterrestrials are ethereal beings, not physical beings. To think extraterrestrials are beyond the physical is the direct result of rejecting science and embracing mythology. Because the non-physical cannot be investigated, not yet at least – do not think that 'ghosts' aren't physical beings, if that's your thing; however, quantum physics has demonstrated, according to Griffin (2017), that they can't exist. This kind of thing has been said before, of course, about many things. Embracing mythology, though, is tantamount to embracing a new religion (or an old one).

Rejecting science results in a diabolical psychological state. The psychological state known as a 'dissociative disorder'. If you'll indulge me, a definition of a dissociative disorder:

Dissociative disorders (DD), according to Wikipedia (2017), are conditions that involve disruptions or breakdowns of memory, awareness, identity, or perception. People with dissociative disorders use dissociation, a defence mechanism, pathologically and involuntarily. Dissociative disorders are thought to primarily be caused by psychological trauma.

'Post-truthers' are an example of people who suffer a dissociative disorder. They reject 'scientific' facts as Truth, and, oddly enough, classify anything they accept as truth (personal truth?) as 'alternative facts'. Alternative facts are not facts. They are symptoms of a pathological condition. Whether their 'condition' was caused by psychological trauma, as claimed by the Wikipedia definition, can only be known by a thorough examination of the individual's psychology. But how many people like this are willing to undergo such an examination? None, that I am aware of. And (perhaps) for good reason: there may

be a stigma attached to willingly submitting to such an examination. No one wants that.

But being unwilling to submit themselves to examination,
'post-truthers', if not ordinary people, suffer a dissociative disorder and raise the risk of being dismissed as untrustworthy.
Dr. J. Allen Hynek decided to accept what people said about
their experience as true (Truth) rather than question what they
said (not question in a negative way, mind you), and based
much of his report, *The UFO Experience: A Scientific Inquiry*,
on their personal reports (even his own) – later referred to as
'experiencers', according to John E. Mack. I am of a different
opinion and, as a researcher, I do not accept what people say
without question, and I think we should question what people
say about UFOs (not in a negative way) and use the scientific
method to investigate their experience (not in a pseudo-scien-
tific approach, but a logical, rational, and systematic one).

Am I suggesting that anyone seeing a UFO undergo a
psych evaluation? No, not at all. What am I suggesting? I'm
suggesting that anyone who claims to have been abducted by
an extraterrestrial make sure they are not suffering some psychological condition. I say this because extraterrestrial contact,
of any kind, would be the most significant event in all human
history. To verify the existence of extraterrestrials would forever alter our perception of ourselves as alone in the universe. So
why not make it easier for science – more so for the sceptics –
by eliminating one possibility? By demonstrating with science
that whoever's had an abduction experience is free from psychological issues would go a long way to demolishing the wall
of disbelief that (some think) has been erected around the subject. I'm thinking of Travis Walton and his experience.

UFO: THE MAKING OF A MYTH?

Refusing to undergo an examination is a right. But refusing to further our understanding of the phenomenon is equal to a witness to a crime refusing to testify. That's how seriously I view such a thing. Experiencers should view it as contributing to the Truth – contributing to developing better scientific ways to understand how, why, and when abduction happens. This is partly why progress has not been made in this area, and for the foreseeable future, won't. Despite recent developments, i.e., the US Navy providing videos that allegedly show UFOs. [And have recently said something echoed in the 1950s: releasing information about UFOs would be detrimental to national security.]

But progress can be made in determining the nature and reality of extraterrestrial craft. And the thus-far-unanswered sixty-four-million-dollar question about whether extraterrestrials exist and are/have been visiting planet Earth (as scientific fact rather than it remain in the realm of speculation).

13. Working hypotheses

Hypothesis: "a supposition or proposed explanation made on the basis of limited evidence as a starting point for further investigation."

A supposition, according to Dictionary.com (2017), is "a belief held without proof or certain knowledge; an assumption or hypothesis." The term 'proposed explanation' is self-explanatory. It is an explanation for something, e.g., the existence of humans is the result of evolution. But it is often conceived because of limited evidence for the thing seen or experienced or known. The existence of humans is a mystery, but one prominent explanation is evolution. Charles Darwin was the first to offer an explanation (based on limited evidence) for the existence of humans and thought it was due to a lengthy process of 'natural selection', which he called evolution. And scientists have sought evidence since then to substantiate his theory. Today, there is more evidence to support the theory of evolution than not. (Compare the theory that genetics not 'survival of the fittest' is the basis for human evolution.)

But what about when we apply this notion to UFOs? The most prominent explanation for UFOs is that they are extraterrestrial craft. Well, this has been the explanation offered since the 1940s at least – based on the idea that humans could not have invented the technology. It is also often revealed in popular culture through books, magazines, and movies – books like Donald Keyhoes' *The Flying Saucers are Real*, magazines like *UFO Magazine*, and movies like *The Day the Earth Stood Still*. However, no serious fundamental research was or has been done to collect data to substantiate the theory that UFOs are extraterrestrial craft. And when I mean serious, I mean the kind that is used to collect data to substantiate the idea that genetics is the bases of evolution. The belief that UFOs are extraterrestrial cannot be used to substantiate the theory that UFOs are extraterrestrial. Collecting evidence does.

The hypotheses that require evidence to substantiate them are 1) UFOs are extraterrestrial technology; 2) UFOs are human technology; and 3) UFOs are misidentifications of a) human technology and b) natural phenomena.

<u>Hypothesis #1</u>: UFOs are extraterrestrial technology.

Is there any evidence to support this? No. However, those who argue 'yes' do so by arguing from five positions on the subject.

1) What was seen couldn't possibly be human technology

2) The witnesses sensory experience couldn't possibly be wrong

3) Contact with the pilots of UFOs, unsolicited (as in abduction or visitation) or solicited (as in CE-5 Contact with Dr. Greer)

4) Argumentation – that is, citing postulations like the Fermi Paradox (Enrico Fermi) or the ET Hypothesis (Jacques Vallée)

5) Speculative Ufology

As I have shown already, the first point is due to ignorance – humans couldn't possibly have invented the technology. Point 2 is due to unquestioning belief in sensory data being accurate – naivety. As I have argued, MUFON has shown that nine-out-of-ten sightings are misidentifications. Point 3 has yet to be verified using scientific methods. Although Dr. Greer's project has much patronage by people wanting to see a UFO or meet an extraterrestrial through solicited contact, his belief that UFOs are extraterrestrial is still unsubstantiated – which begs the question, Who or What is Dr. Greer making contact with? Point 4 is closer to what we might conceive of as the beginnings of actual research. However, the suppositions of Enrico Fermi and Jacques Vallée both need to be proceduralised and data collected. They are arguments derived from or due to the argument that UFOs are extraterrestrial craft because X. Point 5 is almost self-explanatory. This group of people speculate about the nature of UFOs and conclude from their speculation that UFOs are extraterrestrial craft.

The reader can see, or I hope can see, the problematic nature of each of these positions. Points 3 to 5 are baseless positions. By that, I mean that they draw on two things to make judgements about and conclude that UFOs are extraterrestrial. The first source of data is sightings of UFOs, and the second is based on theoretical physics. Jacques Vallée investigated UFOs through sightings (his own included), while Enrico Fermi was addressing the supposition that humans couldn't possibly be alone in the universe (using logic and mathematics). Fermi's supposition looks for planets harbouring intelligent life to confirm the hypothesis, while Vallée's supposition requires two things: 1) a UFO (to verify UFOs are extraterrestrial technology), and 2) an extraterrestrial (to verify UFOs are extraterrestrial craft and confirm Fermi's supposition that we are not alone in the universe).

<u>Hypothesis #2</u>: UFOs are human technology.

MUFON did us a great favour when it established the case management system or CMS. The CMS has amassed a huge database of cases – 80,000-plus as of 2017. Though the cases are sightings of UFOs, this has helped substantiate the original hypothesis about whether UFOs are real. UFOs are real objects. However, the other thing it did was that it didn't substantiate the hypothesis that UFOs are extraterrestrial craft. It has yet to do this, if it ever will. The other thing MUFON did, which I think must annoy the hell out of the believers, is that it showed that nine-out-of-ten UFO sightings were verified as mundane objects, e.g., natural phenomena and human technology. So much for Point 2 above, which argues that sensory experience is perfect when it comes to generating evidence (the only evidence) to substantiate the theory that UFOs are extraterrestrial craft.

Hypothesis #3: UFOs are misidentifications of a) human technology and b) natural phenomena.

MUFON has demonstrated that nine-out-of-ten UFOs are misidentifications. Through excellent field investigative techniques, it has shown that UFOs tend to be misidentifications of a) human technology and b) natural phenomena. However, while MUFON has amassed a database of 80,000-plus cases, it has not provided data to substantiate the theory that UFOs are extraterrestrial. Neither have the people belonging to Points 2 to 5 above. Though they argue UFOs are extraterrestrial, especially the believers in Points 2 and 5, they cannot provide data to substantiate the theory that UFOs are extraterrestrial. At first glance, though many arguments generated by these groups are cogent and logical, they have yet to proceduralise their arguments and collect data to substantiate them.

What is needed for Ufology to become what it desires (and

claims) to be?

Ufology requires what every science requires: A knowledgebase and qualified members. The knowledgebase has yet to be clarified, e.g., UFOs or sightings of UFOs. If UFOs, then Ufology needs one or two objects in its possession, but not in a negative way. Photographs and videos and documents and sightings are the 'limited evidence' that leads to further research. They must be systematically studied like anything else, not excepted on face value, meaning, taken as evidence (a scientific fact) of extraterrestrial craft. This suggests the type of researchers required to generate knowledge in this field (e.g., physicist, engineers, etc.). If sightings of UFOs, MUFON is your resource. This suggests the type of researchers required to generate knowledge in this field (e.g., trained field investigators, ex-police, ex-military [or current members], etc.). And, at the time of writing, only one individual has been granted a doctorate in Ufology, Martin Plowman (2011). However, you may not want to cite him; he concluded UFOs weren't real. This suggests his approach and methodology were faulty. But he was granted his PhD. Go figure.

Ufology requires a registered college or school or university that supports the study of UFOs (or the study of peoples' claims about what they see). The University of Melbourne doesn't really achieve this though it bestowed the first PhD in Ufology. The college should have registered teaching staff and courses of study that lead to certification, diploma, or degrees. If Martin Plowman can gain a PhD, then the potential is there for the college to gain the prestigious accolade of being a research institute in the field of UFO studies. And lastly, the college must publish a peer-reviewed journal; how often is entirely up to the institute. But peer review must be a policy. As well as replicability – replicating the findings of all research by its members, including peer review from other qualified people outside the institute.

No such entity exists at the time of writing. MUFON has the potential to become such an institute, or part of one, if it can pull its head out of the 'unprofessional' hole it has dug for itself. The field of Ufology, and I use the term loosely, is a mess. Who will take up the challenge to organise this wild entity called Ufology?

14. Like a dog chasing its tail

A brief overview of the scene. We have claims of seen objects. We have photographs and video of alleged UFOs – the objects in the photos and videos remain unidentified, at least (they may be real objects, but they could also be models or thrown objects like hubcaps and frisbees – a wink and a nudge). We have sightings of alleged extraterrestrials – a video of an extraterrestrial peeking in someone's kitchen at midnight is not evidence of an extraterrestrial; it could be good CGI or a puppet's reflection in the window. We have claims of being abducted by said extraterrestrials (not the literal puppets, perhaps the metaphorical ones) or of initiating contact. We have professional and 'scientific' studies of UFOs – which, when examined, turn out to be studies of peoples' reports of sightings or of military and government involvement in an investigation of a sighting. But what we do not have is a UFO or an extraterrestrial – and I don't mean in custody for trespassing. And these are the most important elements of all. Without them, nothing gets verified. It's all hearsay and conjecture. Just another 'faith' school.

As I have argued above, no one is studying UFOs, scientifically or otherwise. Unless you believe a secret military group is doing it. The others can't because they can't find a UFO to put under the microscope, as it were. This is an incontestable part of doing science. As would happen in any science of something. This then becomes a dilemma for anyone or any group claiming to be scientifically studying UFOs. UFO studies, like cultural studies or media studies, would prove more authentic a title. Likewise for studying UFOs as extraterrestrial craft. Only one individual seems to have figured out the math for building such a craft, the engine of a UFO, a man called Stan Deyo. Yet his invention is nowhere to be seen.

I have argued above that individuals and groups like MU-FON are in fact studying peoples' claims. MUFON has compiled a huge database of sightings, 80,000-plus (as of 2017), and regularly investigates them using professional methods and techniques. However, not a single investigation has led to the acquisition of or availability of a UFO for analysis. Unless, as MUFON has shown, that it was something mundane. That's because MUFON isn't trying to obtain a UFO; it is merely verifying that a sighting of something is legitimate; meaning, the eyewitness did see something. It, therefore, tries to 'credibilise' sightings. Nothing wrong with that.

What readers may think, after reading this book, is that I'm just another debunker. A sceptic. This is not the case. In fact, I have had many sightings and close encounters of my own. And perhaps I've been abducted, but I have trouble understanding why an extraterrestrial would want to abduct me. I am a trained researcher with a PhD, so I don't accept that we should accept this without question. Accept it all on faith, as some have suggested. Why would we want to? If it is to make people feel good about themselves, fair enough. Why accept without question what may be potentially the greatest moment in human history? If I sound like Dr. Eleanor Ann Arroway in Contact with that last question, it was intentional.

If extraterrestrials are spiritual beings, as many have claimed, then why do they need vehicles to get around in? This is the kind of irrational, illogical thinking that has spurred me to write this book or booklette or whatever you want to call it. Treatise? Why make me look bad by arguing that such is the case? Why be dismissive of the potential for science to help bring the issue to the wider public? Why dismiss the help offered by the few researchers in the scientific community willing to help? Why take an anti-science stance, just because the answers aren't coming fast enough for you? Despite what some may say, impatience is not a virtue.

Many people involved in this issue are like a dog chasing its tail. They can see the tail, but don't register that it belongs to them. So the chase never stops (See TV shows like *Chasing UFOs*). And it won't stop until everyone recognises the futility of maintaining appearances; of not recognising that what they are doing is not what they claim to be doing; that keeping science at arms-length prevents them from moving forward in the pursuit of Truth – and by that, I mean science fact. By embracing belief over fact puts them at odds with the Truth and puts the Truth as far away as Alpha Centauri – even though NASA scientists say that with new technology we may soon get there in two weeks rather than a thousand years.

15. Myth making

When someone believes something is true, without knowing whether it is or not, they give power to a myth. This is how myths are made. Or propagated, according to people like Carl Sagan (Quotes, 2017), Myth (2017), and Mythopoeia (2017). Until they're introduced to new information that dispels the myth and their belief is modified or changed. There are two meanings of the word myth. Firstly, according to Dictionary.com, myth is "... a traditional story, especially one concerning the early history of a people or explaining a natural or social phenomenon, and typically involving supernatural beings or events." And the second, "a widely held but false belief or idea (Ibid.)." If UFOs are seen as 'magic' or 'supernatural beings', then the first meaning applies. However, if UFOs aren't real, then the second meaning applies.

I have argued or shown or agreed with the facts that UFOs are real objects. This is not being questioned. Therefore, UFOs are not a 'false belief or idea'. There is plenty of evidence to show they are real. So they are not a myth. And this evidence is no longer disputed. Sightings of things in the sky have been documented and confirmed since the first reports recorded. MUFON has, as of 2017, 80,000-plus cases. The question isn't 'Why are people seeing things?' because that question leads to a black hole where the information gets scrambled. The question is, 'What are they seeing?'

MUFON has shown that nine-out-of-ten cases are misidentifications. If so, why are people claiming they are not? Mostly because they see what they want to see. I mean, if they think they are seeing something extraterrestrial, then that is what they think they see. They continue to think it despite the evidence to the contrary – which means most of the evidence. Religion creates the same psychological phenomenon. Despite evidence the 'deity' believed in may not exist, and, as it turns out, a mythological entity.

When people give ascent to stories about 'deities', they give power to a myth. Now don't get me wrong here. If believing in a deity makes that person a 'better' person, and I mean psychologically and socially (physically tends to follow from good psychology, e.g., regular health checks, self-monitoring, vaccinations) – unlike ISIS people, for instance – then that is a good thing. But if believing makes you do 'bad' things, like create hoaxes, deceive people, mislead them, or down-right lie to them, perhaps even kill them, according to Austin (2016), then believing hasn't done you any favours. This is called ethics.

Believing something because it makes you feel good about yourself is one thing. But believing something because you think it is reasonable and logical, is quite another. Believing that UFOs are real is reasonable, given the evidence that supports their existence – e.g., government and military investigations, MUFON field investigations, civil aviation pilots and radar operator reports. However, believing UFOs are extraterrestrial is yet another thing. And believing they are pushes you along the path to giving power to a myth, until the evidence is in. Why? Because there isn't any scientific evidence (as yet) to support the idea that UFOs are extraterrestrial.

And therein lies the problem.

Where is the scientific evidence to support the claim that UFOs are extraterrestrial? The ideas that give credence to the belief that UFOs are extraterrestrial are 1) their 'magical' appearance, and 2) the seemingly advanced technology. We've discussed both these ideas, and neither is logical. Magic is attributable to a naïve or ignorant mindset. Only naïve people think things are 'magical' because they can't explain them, or the thing defies logic. And ignorant people say humans can't have been smart enough to invent the thing. It is irrational to think extraterrestrials wouldn't also invent the technology.

Whatever is happening to people who believe they are seeing extraterrestrials begs to be thoroughly investigated. And I don't mean half-heartedly using pseudoscientific approaches. If extraterrestrials are visiting planet Earth, then we need to know about it. We need to know why and how, not just when. And that requires a thoughtful, serious, systematic approach. Not merely a psychological one, the likes of which John E. Mack (1999) conducted in his PEER – Program for Extraordinary Experience Research at the John E. Mack Institute (2003). Others, ICAR: International Center for Abduction Research, have also done this type of research.

John Mack's research suggests that this type of extraterrestrial 'visitation' is diabolical. It isn't friendly at all. Invading someone's mind is diabolical. It is unethical. Unless it was invited. But why invite it? Why open your mind to something that is willing to invade you and give you nightmares? Perhaps people have been influenced by the Neo-Christian doctrine that says the Holy Spirit must live within you to receive Jesus Christ's blessings and took it literally. You invite the Holy Spirit in and *voila*! you're a 'born again' Christian. It's a 'magical' event, but it's real to the person experiencing it.

If we rely on such stories to give us focus, then we display a naivety of grand proportions. I'd like to throw in a joke about gullibility here, but it may not get the desired response. Some people are perfectly happy with this state of affairs – inviting 'spirits' to live in them. But this is the stuff of mythology. And such people would prefer to believe these things than believe science, even though science has given them everything (today) and perhaps even the life they now live. Such a mindset is indicative of a dissociative disorder, which I've discussed.

So far, believing UFOs are extraterrestrial in origin is equal to giving power to a myth. There is no scientific evidence to support the idea. Only conjecture. But does that mean we should not think it's possible? No. On the contrary, we should be thinking up ways (methods) to verify it. It has the potential to change humanity, in profound ways, according to Richard Dolan and Bryce Zabal, in their book called *A. D. After Disclosure*. There are plenty of good reasons to think it's possible. Astrophysics is discovering new planets every day. Astrobiology notes that some living things can exist in space or not be killed, e.g., Tardigrades. If such things can exist off-world, then they can exist elsewhere. And perhaps, being speculative, everything came from elsewhere before it existed on Earth. A reasonable hypothesis, given the discoveries of the past ten or twenty years. Aside from quantum physics, that is.

Yet, some myths persist.

Perhaps this is what Karl Marx meant when he quipped that 'religion is the opium of the people'. If it gives you focus, in a good way, then it is unproblematic.

And that is where I'll leave mythmaking.

Predicting the future is relatively easy when the tail you chase is something implicit in yourself. If your capacity to deceive yourself is easy, then there's no cure for it. As evidenced by the 'post-truther' movement. Unless you accept new evidence that contradicts your belief – or new evidence to revise your hypothesis. If you believe something is fact though it isn't, you deceive yourself – sometimes in a good way, sometimes not. When you invite others to believe your 'alternative' facts, you cross the line of what is considered ethical behaviour, I think. It is but a short distance from there to commit a crime. By this I mean, knowing you're short on the evidence (and don't admit it) and promote your belief as fact (when it's merely a personal belief), you may end up with more than egg on your face.

The link between UFOs and extraterrestrials is missing, and we must find it. And that's pretty much all I've been saying. It's a moral imperative, now, given what has happened recently – the advent of new evidence (e.g., the US Navy videos, the US Congress passing a Bill to legally establish a group to properly investigate and report back to the government on UFOs [I wish them luck!]), that substantiates the theory that UFOs are real, should make a person revise their theory. We now need the evidence that links UFOs with extraterrestrials. Otherwise, Ufology and Ufologists will forever live in Plato's cave – see Plato's 'allegory of the cave'. The members of this loose-knit, messy community called 'Ufology' are presently the prisoners chained to the wall of Plato's cave, giving names to the shadows that pass before them on the wall. And no desire to escape the cave. They must break those bonds and leave. Get out and find the missing link, which some Ufologists are trying to do, e.g., Richard Dolan. We also must be like Heraclitus in Lewes' (1845) *A Biographical History of Philosophy*, denouncing perpetual delusion, but not in the realm of the senses as Xeno-

phanes did, but in the realm of imperfect reasoning.
"As all energy longs for a fitting theatre on which to play its part."
– Ascribed to Parmenides in G. H. Lewes' *A Biographical History of Philosophy.*

Note to the reader

If you are concerned or wish to see a complete list of references for the above work, please send the author a message: easterbrookrm at gmail dot com.

About the Author

A little bit about the author, R. M. Easterbrook. He lives in Canberra, Australia. He sees himself as a global citizen rather than an 'Australian'. In fact, he doesn't even view himself as an Australian. He prefers to think of himself as a British Australian. As a British Australian, he is ashamed of what the British did, and subsequent governments and citizens of Australia have done to the Indigenous of this country since Invasion Day. As an individual, he feels the least he can do is acknowledge the many Indigenous cultures of 'the no name land', of which there were at least two hundred distinct cultures when the British 'invaded' (and now there may be as little as twenty left), and agrees with others that the present 'custodians', as they like to call themselves, have done a shit job of acknowledging these things and honouring these peoples.

He has seen and experienced 'high strangeness' since he was seven or eight. Late one night his family was returning home from spending time in Melbourne (during the holiday season) when a green-glowing object landed about one hundred metres from the road – his siblings were asleep at the time, and his mother, who appeared to be disturbed by the sight of the object more than he was (in fact, he felt strangely calm about it), told him to get down on the floor and go back to sleep. He ignored her entreaty and, small eyes peeking above the ledge of the window, watched the object for as long as possible until it was out of sight. It wasn't the thing that got him into 'Ufology'. The thing that got him into 'Ufology' was the continued sightings, which have occurred regularly since that strange night.

When he was about seventeen, he was given a copy of Stan Deyo's *Cosmic Conspiracy*. The gift-giver thought Deyo's book would enlighten him to what was really going on, but all it did was raise more questions. Questions that have yet to deliver reasonable answers. And the questioning continues. The more education he received, the harder it was to think rationally about UFOs and what they may be. As Richard Dolan would say, mainstream education doesn't say a goddamned thing about the phenomena. And R. M. Easterbrook has received a lot of education and training. Even while completing a PhD (2009 to 2014), the 'elites' in academia said, when he raised the issue, that it was a career destroyer. Don't go there; don't even think about it.

He has engaged in an on/off relationship with the subject of UFOs since his pre-teens. And after thinking about it or being actively involved, at one level or another, and taking the time to look at the phenomenon seriously, he would say he was as much a Ufologist as the next person. Although Ufology (which needs a better name) may be an unrecognised field of study in mainstream higher education (which is ironic and puzzling because a Melbourne University student managed to gain the first PhD in Ufology, and may be the last), is, in his opinion, a multi-disciplinary subject. It draws on the knowledge and experience of diverse fields of study and not just one field, and now (because he has not noticed that anyone from Applied Linguistics, the field of study in which he managed to gain a PhD, has contributed to the discussion) adds his two cents worth with this little booklette of commentary on the field of Ufology. May it serve its purpose, which was to comment on the study of UFOs and the field of Ufology.